D0845627

Infection, Polymorphism and Evolution

Infection, Polymorphism and Evolution

Edited by

W.D. Hamilton

Department of Zoology
University of Oxford
South Parks Road
Oxford OX1 3PS
UK

and

J.C. Howard

Institute of Genetics
University of Cologne
Zuelpicher Strasse
D-50674 Cologne
Germany

CHAPMAN & HALL
London · Weinheim · New York · Tokyo · Melbourne · Madras

Published by Chapman & Hall, 2–6 Boundary Row, London SE1 8HN, UK

Chapman & Hall, 2–6 Boundary Row, London SE1 8HN, UK

Chapman & Hall GmbH, Pappelallee 3, 69469 Weinheim, Germany

Chapman & Hall USA, 115 Fifth Avenue, New York, NY 10003, USA

Chapman & Hall Japan, ITP-Japan, Kyowa Building, 3F, 2-2-1 Hirakawacho, Chiyoda-ku, Tokyo 102, Japan

Chapman & Hall Australia, 102 Dodds Street, South Melbourne, Victoria 3205, Australia

Chapman & Hall India, R. Seshadri, 32 Second Main Road, CIT East, Madras 600 035, India

First edition 1997

© 1997 The Royal Society

Printed in Great Britain by St Edmundsbury Press, Suffolk

ISBN 0 412 63700 6

A catalogue record for this book is available from the British Library

∞ Printed on permanent acid-free text paper, manufactured in accordance with ANSI/NISO Z39.48-1992 and ANSI/NISO Z39.48-1984 (Permanence of Paper).

Contents

Contributors

Michael Aidoo　Molecular Immunology Group, Institute of Molecular Medicine, University of Oxford, John Radcliffe Hospital, Oxford OX3 9DU, UK

Catherine E.M. Allsopp　Molecular Immunology Group, Institute of Molecular Medicine, University of Oxford, John Radcliffe Hospital, Oxford OX3 9DU, UK

M.S. Andersson　Department of Zoology, Uppsala University, Villavägen 9, S-752 36 Uppsala, Sweden

Ian R. Crute　Horticulture Research International, Wellesbourne, Warwick CV35 9EF, UK

Miles Davenport　Molecular Immunology Group, Institute of Molecular Medicine, University of Oxford, John Radcliffe Hospital, Oxford OX3 9DU, UK

Paul W. Ewald　Department of Biology, Amherst College, Amherst, MA 01002-5000, USA

Steven A. Frank　Department of Ecology and Evolutionary Biology, University of California, Irvine, CA 92717, USA

Sarah C. Gilbert　Molecular Immunology Group, Institute of Molecular Medicine, University of Oxford, John Radcliffe Hospital, Oxford OX3 9DU, UK

Sunetra Gupta　Department of Zoology, University of Oxford, South Parks Road, Oxford OX1 3PS, UK

L. Gustafsson　Department of Zoology, Uppsala University, Villavägen 9, S-752 36 Uppsala, Sweden

Adrian V.S. Hill　Molecular Immunology Group, Institute of Molecular Medicine, University of Oxford, John Radcliffe Hospital, Oxford OX3 9DU, UK

R. Stephen Howard　Department of Biology, Indiana University, Bloomington, IN 47405, USA

Carina Y. Howell　Department of Biology and Institute of Molecular Evolutionary Genetics, The Pennsylvania State University, University Park, PA 16802, USA

Austin L. Hughes　Department of Biology and Institute of Molecular Evolutionary Genetics, The Pennsylvania State University, University Park, PA 16802, USA

Marianne K. Hughes　Department of Biology and Institute of Molecular Evolutionary Genetics, The Pennsylvania State University, University Park, PA 16802, USA

Steven E. Kelley　Department of Biology, Emory University, 1510 Clifton Road, Atlanta, GA 30322, USA

Jan Klein　Max-Planck-Institut für Biologie, Abteilung Immungenetik, Corrensstr. 42, D-72076 Tübingen, Germany, and Department of Microbiology and Immunology, University of Miami School of Medicine, Miami, FL 33136, USA

Ajit Lalvani　Molecular Immunology Group, Institute of Molecular Medicine, University of Oxford, John Radcliffe Hospital, Oxford OX3 9DU, UK

Curtis M. Lively　Department of Biology, Indiana University, Bloomington, IN 47405, USA

C. Jo Manning   Center for Mammalian Genetics and Department of Pathology, University of Florida, Gainesville, FL 32610, USA, and Department of Psychology, University of Washington, Seattle, WA 98195, USA

Masatoshi Nei   Department of Biology and Institute of Molecular Evolutionary Genetics, The Pennsylvania State University, University Park, PA 16802, USA

D. Nordling   Department of Zoology, Uppsala University, Villavägen 9, S-752 36 Uppsala, Sweden

Colm O'Huigin   Max-Planck-Institut für Biologie, Abteilung Immungenetik, Corrensstr. 42, D-72076 Tübingen, Germany

Magdalena Plebanski   Molecular Immunology Group, Institute of Molecular Medicine, University of Oxford, John Radcliffe Hospital, Oxford OX3 9DU, UK

Wayne K. Potts   Center for Mammalian Genetics and Department of Pathology, University of Florida, Gainesville, FL 32610, USA, and Department of Psychology, University of Washington, Seattle, WA 98195, USA

A. Qvarnström   Department of Zoology, Uppsala University, Villavägen 9, S-752 36 Uppsala, Sweden

Paul Schmid-Hempel   ETH Zürich, Experimental Zoology, ETH-Zentrum NW, CH-8092 Zürich, Switzerland

B.C. Sheldon   Department of Zoology, Uppsala University, Villavägen 9, S-752 36 Uppsala, Sweden

Edward K. Wakeland   Center for Mammalian Genetics and Department of Pathology, University of Florida, Gainesville, FL 32610, USA, and Department of Psychology, University of Washington, Seattle, WA 98195, USA

Claus Wedekind   Abteilung Verhaltensökologie, Zoologisches Institut, Universität Bern, CH-3032 Hinterkappelen, Switzerland

Simon N.R. Yates   Molecular Immunology Group, Institute of Molecular Medicine, University of Oxford, John Radcliffe Hospital, Oxford OX3 9DU, UK

Preface

The peculiar capacity of infectious disease pathogens to drive polymorphism into host resistance systems was first recognized by J.B.S. Haldane nearly 50 years ago. Haldane's insight was to recognize that relatively high rates of evolution by pathogens prevent the achievement of static optima for successful resistance systems in the host population. On the other hand, the necessities of pathogen transmission guarantee a period of relative selective advantage for resistant host variants as long as they are too rare to attract the pathogen's attention. The outcome of such a relationship is the evolution of a host population where individual resistance genotypes are, on average, rare and associated with periodic fluctuations in the frequency of resistant alleles in the host population and of complementary virulent forms among pathogens. The existence of exceptionally polymorphic resistance loci such as the major histocompatibility complex (MHC) of vertebrates or the universality of population fine structure among infectious pathogens finds a satisfying explanation in principle in the light of this idea. Haldane pointed out that the consequence of frequency-dependence of successful resistance to infectious parasites should be that optimal mutation rates for resistant loci should be characteristically high. However, genotypic rarity is also achieved by the combinatorial reshuffling of polymorphic loci and the association of different allele combinations at a locus by the sexual process, and one of the revelations of the recent period has been that parasite-driven selection is easily capable of sustaining sexual reproduction in a population by this means, despite the cost of sex. If resistance to infectious disease is indeed a key factor in the origin and maintenance of sex, then many of the social and phenotypic manifestations of sexual reproduction may have to be re-evaluated in the light of this idea.

Our meeting at the Royal Society was designed to bring together a number of the strands of this complex field. Theoretical and field work on reproductive and behavioural strategies in the face of pressure from infectious pathogens were put together with genetic, biochemical and epidemiological work on the major histocompatibility complex of vertebrates. The high level of generality of the issues meant that studies on vertebrates could be usefully discussed in the same context as work on invertebrates and plants. Perhaps the only major aspect of this field that was underrepresented at the meeting was the study of pathogen strategies, although this was compensated for to a certain extent at the time by Professor Keith Leewenhoek's Lecture entitled 'The Opportunistic Pathogen' which took place during the meeting.

R.C. Howard
Cologne, Germany
March 1996

1

Selection by parasites for clonal diversity and mixed mating

CURTIS M. LIVELY AND R. STEPHEN HOWARD*

Department of Biology, Indiana University, Bloomington, Indiana 47405, U.S.A.

SUMMARY

On theoretical grounds, coevolutionary interactions with parasites can select for cross-fertilization, even when there is a twofold advantage gained by reproducing through uniparental means. The suspected advantage of cross-fertilization stems from the production of genetically rare offspring, which are expected to be more likely to escape infection by coevolving enemies. In the present study, we consider the effects that parasites have on parthenogenetic mutants in obligately sexual, dioecious populations. Computer simulations show that repeated mutation to parthenogenesis can lead to the accumulation of clones with different resistance genotypes, and that a moderately diverse set of clones could competitively exclude the ancestral sexual subpopulation. The simulations also show that, when there are reasonable rates of deleterious mutation, Muller's ratchet combined with coevolutionary interactions with parasites can lead to the evolutionary stability of cross-fertilization. In addition, we consider the effects that parasites can have on the evolution of uniparental reproduction in cosexual populations. Strategy models show that parasites and inbreeding depression could interact to select for evolutionarily stable reproductive strategies that involve mixtures of selfed and outcrossed progeny.

1. INTRODUCTION

Consider two obligately asexual clones, A and B, which are alike in all ways, except that clone A produces twice the number of daughters. If these two clones were to be placed in direct competition, clone A would eliminate clone B in very few generations. Now consider a similar competition between clone A and a sexual group. Both groups produce the same number of offspring, but only half of the offspring produced by the sexual individuals are daughters. Hence the production of daughters by the sexual group is the same as that for clone B above, and clone A is again expected to win. Why then is there sex (Maynard Smith 1971)? Any sexual population that generates a rare asexual mutant should be rapidly replaced by the descendants of that mutant, unless the clone actually produces fewer daughters.

The reasoning makes sense, but the predominance of sexual reproduction in eukaryotes suggests that there is more to the outcome than simple daughter production. The 'alike-in-all-ways' assumption made above is apparently false. One way in which the assumption is false is that clonal populations cannot reduce their mutational load by concentrating mutations in a few low-fitness offspring (Muller 1964). Because of the probabilistic nature of muta-

tion, Muller viewed an asexual population as a monophyletic series of subclones, each having a different number of deleterious mutations. He reasoned that, by chance alone, all the members of the subclone with the lightest mutational load could fail to reproduce. Alternatively, by chance, all the offspring from the members of the least loaded clone could acquire an additional mutation. Using this kind of reasoning, Muller foresaw an 'irreversible ratchet mechanism' leading to an ever-increasing number of deleterious mutations in clonal lineages, especially in small clonal populations (Muller 1964). Some striking examples for the operation of Muller's ratchet are given by Leslie & Vrijenhoek (1980), Chao (1990), Bell (1988) and Rice (1994).

Another way in which the 'alike-in-all-ways' assumption is false is that clonal populations lack the genetic variability of sexual populations, and they therefore lack the same potential to respond to selection (Weismann 1989). Here we are mainly concerned with the second alternative, that sexual populations have an advantage that depends on their ability to produce variable progeny. Specifically, we are concerned with the recent idea that the production of variable progeny is advantageous as a defence against parasites (Levin 1975; Jaenike 1978; Glesener & Tilman 1978; Hamilton 1980, 1982; Bremermann 1980; Lloyd 1980). In its simplest form, the idea is very straightforward. Any clone will have an advantage when rare, because of its enhanced

* Present address: Department of Biology, Middle Tennessee State University, Murfreesboro, Tennessee 37132, U.S.A.

daughter production. However, as that clone becomes common, there is an increasing selection on its parasites to be able to infect it. This kind of frequency-dependent selection against the clone (and common genotypes in general) could prevent the clone from replacing its sexual ancestor.

The idea, now known as the Red Queen hypothesis (after Bell 1982), has appeal for two reasons. One is that parasites seem ubiquitous; there are even parasites of parasites. The other reason is that the ecological correlates of biparental reproduction are consistent with the Red Queen hypothesis. Sexual species tend to inhabit undisturbed, biologically complex habitats where parasitism is expected to be common (Levin 1975; Glesener & Tilman 1978; Bell 1982). In addition, recent studies have shown that cross-fertilization predominates within populations of the same species where parasites are common, and that uniparental modes of reproduction predominate in populations where parasites are rare (Lively 1987, 1992; Schrag *et al.* 1994*a*). Explaining the ecological distribution of cross-fertilization is a necessary requirement for any general theory of sex.

The most critical assumption of the parasite (or Red Queen) theory of sex is that there is increasing selection against host genotypes as they become common, independent of whether they are sexually or asexually produced. This selection imposed by parasites sets up the expected time-lagged oscillations in host and parasite gene frequencies from which the Red Queen hypothesis gets its name (Hutson & Law 1981; Bell 1982; Bell & Maynard Smith 1987; Nee 1989). In what follows, we present the results of computer simulations of host–parasite coevolution which suggest that time-lagged frequency-dependent selection may select for clonal diversity, and that sexual reproduction can be replaced by a moderately diverse set of clones. We also present strategy models for the evolution of selfing in cosexual populations. The results of these models suggest that parasites can select for mixtures of selfed and outcrossed progeny within a single brood.

2. PARASITES AND CLONAL DIVERSITY

Consider the spread of an initially rare apomictic clone in a dioecious sexual population. As indicated above, the clone should spread when rare, owing to its enhanced daughter production. Then, as the clone becomes common, it is expected to be 'tracked' by parasites, and driven down in frequency if the effects of parasites on host fitness are sufficiently severe. However, unless the clone is driven extinct, it should recover its original rare advantage and begin increasing again. Under these conditions, the outcome of repeated mutation to clonal reproduction could lead to the accumulation of clonal diversity, which would diminish the advantage of cross-fertilization. If, however, clones accumulate deleterious mutations, then the twofold advantage of parthenogenetic reproduction will be eroded during the initial spread of the clonal mutant, and each time the clone is driven down in frequency by the parasite

(Howard & Lively 1994). Here we evaluate the effects of parasites on accumulation of clonal diversity under two conditions: (i) without recurrent deleterious mutations; and (ii) with an average of one deleterious mutation per genome per generation.

(a) Methods

To address this question, we employed an individual-based computer simulation of host–parasite interactions. Hosts and parasites were treated as haploid individuals whose interactions were mediated by two unlinked diallelic loci. All hosts contained an additional 500 unlinked loci at which harmful mutations could accumulate with a probability of $U/500$ per generation, where U is the Poisson-distributed mean mutation rate per genome per generation. The host population ($N = 1000$) also received a sexual migrant on average once every two generations. These migrants were randomly assigned one of the four possible parasite-compatibility genotypes, and were initialized with a Poisson-distributed mean of U/s deleterious mutations, where s is the selection against deleterious mutations (Maynard Smith 1978).

Parasites were obligately sexual and reproduced twice for each bout of host reproduction. To maintain genetic variation in the parasite population, alleles at the interaction loci mutated to the alternative state with a mean probability of 0.03. At the beginning of each parasite generation, individual hosts were drawn sequentially from an array and matched against a randomly drawn parasite with probability (T). A parasite was successful at infecting a host if, and only if, an exact genetic match occurred at both interaction loci. Parasites that were successful at infecting hosts were placed in an array, and allowed to participate in a reproductive lottery at the end of each parasite generation; unsuccessful parasites died. Hosts were marked as either parasitized or unparasitized and returned to the source array; each host could be infected by a maximum of one parasite.

At the end of each host generation, individuals were selected randomly for reproduction. If the chosen host individual was sexual, a second host was selected at random to serve as a mate. Sexual pairs mated and gave rise to haploid embryos through a process analogous to zygotic meiosis; asexual females reproduced mitotically. The number of offspring produced by hosts was discounted according to the effects of parasitism (E) and selection against deleterious recurrent mutations. Viability of offspring in the presence of deleterious mutations was calculated as $(1 - s)^k$, where s is the selection coefficient and k is the number of mutations (after Maynard Smith 1978); here we set $s = 0.025$. Host reproduction was allowed to continue until the total number of broods produced equalled the number of adults in the population. Surviving embryos were then selected at random without replacement to become the next generation of adults. To simulate repeated mutation to parthenogenesis, a clonal mutant with a twofold reproductive advantage was randomly selected from the sexual population on average once every 60 host generations. Clones derived in this fashion were identical to their sexual parent with respect to the genetic configuration at their parasite interaction loci and the number of harmful mutations in their genomes. (Copies of the computer code are available on request.)

We conducted the initial runs of the model in a large region of parameter space in which we previously observed coexistence between a sexual population and a single clonal mutant (Howard & Lively 1994). Specifically, we initialized the parameters at $T = 0.90$ (90% risk of contact with a randomly drawn parasite), $E = 0.70$ (70% reduction in the fitness of infected hosts), and $U = 0$ (i.e. no deleterious mutation). The purpose here was to determine the number of clones needed to displace the sexual population in the absence of deleterious mutation. We then added deleterious mutations to the model at the rate of one per genome per generation. This rate of mutation has empirical support from studies on *Drosophila* (Mukai *et al.* 1972; Houle *et al.* 1992). Finally, we added a third diallelic locus to the model for both the host and the parasite.

(b) Results

In the absence of deleterious mutation ($U = 0$), a sexual population with four different resistance genotypes (two diallelic loci) was eliminated soon after invasion by only the second clonal host genotype. The mean time to elimination of sex was 27 generations ($N = 200$) after entry into the population by the second clone. A representative run is given in figure 1*a*. In this run, a randomly sampled clone was 'spun off' at generation zero, and it quickly began to oscillate with the sexual population. Then at generation 19, a second clone was randomly sampled from the sexual population, and it increased rapidly. By generation 80, the sexual population had been driven extinct, and the two clones continued in a stable oscillation with each other. The same basic result was observed in the three-locus model. In 200 runs of the simulation, the sexual population was driven extinct an average of 36 generations after the second clone entered the populations (a representative run is given in figure 1*b*). Hence, it seems that parasites by themselves are insufficient to provide protection for sex in the face of repeated mutations to parthenogenetic reproduction. The results also suggest that the diversity of resistance genotypes in a clonal population need not equal the diversity of the sexual population in order for clones to drive sexual populations to extinction.

A different result was gained when individuals gained an average of one deleterious mutation per generation ($U = 1$). In both the two-locus and the three-locus models, the sexual population persisted until the run was terminated at 5000 generations, even though a randomly sampled clonal mutant entered the sexual population every 60 generations, on average. Particularly interesting parts of these runs are presented in figure 2, which shows the successive invasion and extinction of clones having different

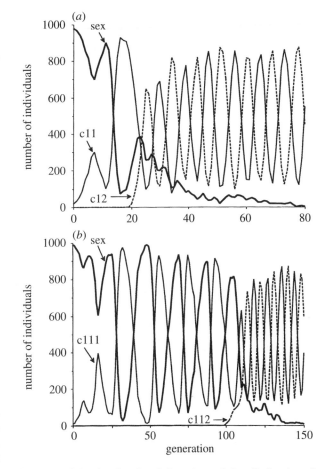

Figure 1. Selection for clonal diversity and the elimination of sex under parasitism in the absence of mutation accumulation. The heavy line represents the population trajectory of a freely recombining sexual population in competition with one, and then two, asexual genotypes. In (*a*), the two-locus model, each asexual lineage was initialized with one of the four possible parasite-compatibility genotypes extant in the sexual population (labelled c11, c12, c21, c22); in (*b*), the three-locus model, each asexual lineage was initialized with one of eight possible parasite-compatibility genotypes extant in the sexual population (labelled c111, c112...c222). Parameters for these runs of the simulation included a probability of parasite transmission (T) = 0.9 and an effect of parasites on host fitness (E) = 0.7 (i.e. an average of 70% reduction in host fecundity as a result of parasitism).

genotypes for parasite resistance. In all four cases, the clones oscillate several times before undergoing a mutational meltdown (in the sense of Lynch & Gabriel 1990). The meltdown is due to the accumulation of mutation through Muller's ratchet, which is aided by the parasite-driven oscillations in populations size of the clonal lineages (see Howard & Lively 1994).

We then reduced the effects of parasites from $E = 0.70$ to $E = 0.50$ and held all other parameters constant. The simulated sexual population again persisted until the run was terminated at 5000 generations in the two-locus model. Because of the reduced effect of infection, the clones persisted for a longer time; this persistence increased the likelihood that they would overlap in time. Figure 3*a* shows a part of the run in which three separate clones with different parasite-resistance genotypes coexisted for

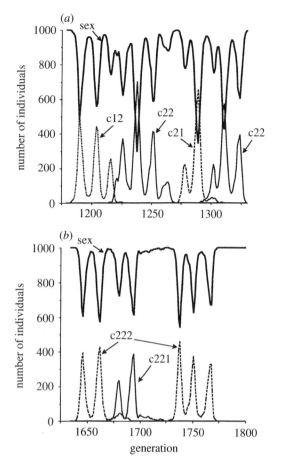

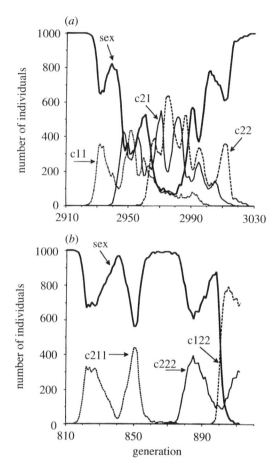

Figure 2. Elimination of asexual genotypes by the combined effects of parasitism and mutation accumulation. The heavy line labelled 'sex' represents the population trajectory of a freely recombining sexual population undergoing repeated mutation to parthenogenesis; asexuals were derived from the sexual population at an average rate of once every 60 generations. In (a), the two-locus model, the asexual lineages represented random samples of genetic variation in the sexual population with respect to both numbers of deleterious mutations and parasite compatibility genotypes (c11, c12, c21, c22). In (b), the three-locus model, the conditions were the same except that there were eight possible parasite compatibility genotypes (c111, c112...c222) that caused infection when they matched a similar eight possible host genotypes. Parameters for this run of the simulation included a per genome mutation rate (U) of 1.0, an effect of mutation $(s) = 0.025$, a probability of parasite transmission $(T) = 0.9$, and an effect of parasites on host fitness $(E) = 0.7$ (i.e. an average of 70% reduction in host fecundity as a result of parasitism).

over 40 generations. All three clones eventually went extinct, owing to the combined effects of parasites and mutation accumulation. However, in the three-locus model, the sexual population was eliminated by a pair of clones at generation 910.

Finally, we reduced the effects of parasites to $E = 0.40$. In this simulation, a pair of clones drove the sexual population extinct in less than 2400 generations in both models (figure 4). Hence, under the conditions of our simulations, moderately severe effects of parasites $(E > 0.50)$ were required to protect the sexual population from replacement by multiple clones. The minimum effects of parasites needed to

Figure 3. Competition between sexual and clonal populations facing the combined effects of parasitism and mutation accumulation. The parameter values are the same as in figure 2, except that the effects of parasites $(E) = 0.5$. (a) Two-locus model; (b) three-locus model, as described in figure 2. Note that the sexual population is driven extinct in the three-locus model.

stabilize sexual populations, however, may be reduced in populations larger than simulated here $(N = 1000)$. This may be expected because, in larger populations, clonal lineages will have more time to accumulate deleterious mutations, which in combination with parasites helps to erode their fecundity advantage before they replace the sexual population.

3. PARASITES AND MIXED MATING

Nested within the question 'why cross-fertilize?' is a more subtle question: why partial cross-fertilization? Many organisms that produce cross-fertilized progeny also produce uniparental progeny in some form. Sponges, for example, produce outcrossed progeny, but they also produce gemmules, which are vegetatively produced, physiologically independent offspring. Bryozoans similarly produce mixtures of uniparental and biparental offspring, as do many other invertebrates and plants (reviews in Bell 1982; Bierzychudek 1987). Of particular interest is self-fertilization, which is a uniparental sexual process. The question is: why mix it up? And what is the optimal mixture of uniparental and biparental offspring? In what follows, we present models of

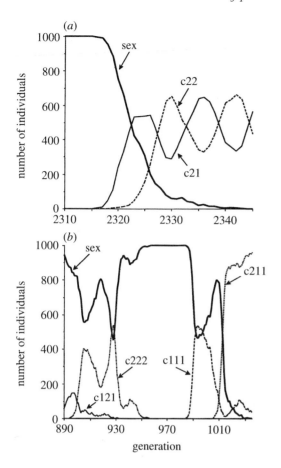

Figure 4. Selection for clonal diversity and the elimination of sex under parasitism in the presence of mutation accumulation. The parameter values are the same as in figures 2 and 3, except that the effects of parasites $(E) = 0.4$. (*a*) Two-locus model; (*b*) three-locus model, as described in figure 2. Note that the sexual population is driven extinct in both models. Compare with figure 1.

selection on mating systems when there are parasites and pathogens to contend with. The results suggest that mixed reproduction can be evolutionarily stable over a wide range of conditions, depending on how the risk of parasitism is associated with increasing levels of selfing in the population.

(*a*) *A strategy model*

Consider a simultaneous hermaphrodite in a large, randomly mating population. Assuming that reproduction through female function is resource limited, the expected fitness of this individual (W_i) can be approximated by summing up the fitness gains through male function (W_m), plus female function through outcrossing (W_f), plus female function through uniparental (parthenogenetic or self-fertilized) ova (W_p). Thus

$$W_i = W_m + W_f + W_p, \tag{1}$$

where:

$$W_m = a_i r_m V, \tag{2a}$$

$$W_f = (1 - a_i)(1 - p_i) r_f, \tag{2b}$$

$$W_p = (1 - a_i) p_i y. \tag{2c}$$

In this formulation, a_i is the individual's allocation to male function, and a is the population mean; p_i is the individual's allocation to uniparental ova, and p is the population mean; y is the fitness of uniparental ova, relative to fully outcrossed ova; r_f is the percentage of the genome passed onto amphimictic zygotes through female function, and r_m is the percentage of the genome passed onto amphimictic zygotes through male function ($r_f + r_m = 1$). Finally, the reproductive value of male function, V, is equal to $(1 - a)(1 - p)/a$. This formulation for W_i assumes that the use of male gametes for the purpose of self-fertilization does not affect the total number of outcrossed progeny gained through male function (i.e. no 'pollen discounting' in the sense of Holsinger *et al.* 1984). It also assumes that all ova are fertilized, and that uniparental ova are not sequestered in cleistogamous flowers (see Lively & Lloyd 1990).

Consider the situation where the fitness of uniparental progeny, y, depends on three variables: (i) inbreeding depression (d), or, equivalently, the effects of developmental defects in parthenogenetic progeny; (ii) the effects of parasites (E); and (iii) the rate of parasite contact or transmission (T), as follows:

$$y = (1 - d)(1 - ET). \tag{3}$$

Here, we let the transmission rate, T, be a function of the average allocation to uniparental ova within the population (p). Specifically, we set the contact or transmission rate as

$$T = \left[\frac{(N - 1)p + p_i}{N} \right]^z, \tag{4}$$

where N is the number of individuals in the population. In effect, we assume that the probability of contact with parasites approaches unity as p and p_i approach one. The rationale for this assumption is that, as selfing or parthenogenesis increases in the population, genetic diversity decreases, thereby facilitating the spread of pathogens. The exponent (z) controls the shape of the transmission function, and is expected to depend on the type of uniparental progeny, the aggregation of siblings, the genetic basis for disease resistance, and the type of parasite or pathogen. The case $z > 1$ means an initially low but accumulating effect of increasing p on the probability of contact with parasites, whereas $z < 1$ means an initially high but decelerating effect (cf. 'peaky' and 'pitty' fitness profiles in Hamilton (1993)).

The effect of small changes in allocation to uniparental ova on individual fitness can be calculated by taking the first partial derivative of W_i with respect to p_i. At the evolutionarily stable strategy (ESS) (Maynard Smith 1982) or unbeatable strategy (Hamilton 1967), the population mean (p) is equal to the optimal allocation to uniparental progeny for the individual (p_i). Hence the derivative can be solved for $p_i = p$. Moreover, at the ESS, the derivative is equal to zero. Setting the derivative equal to zero and solving for the equilibrium value of $p(p^*)$ gives:

$$p^* = \left[\frac{1 - d - r_f}{E(1 - d)(1 + 1/N)}\right]^{1/z}. \qquad (5)$$

Assuming that the number of individuals in the population (N) is large, and substituting one half for the relatedness to offspring through female function, r_f, into equation (5) gives:

$$p^* \approx \left[\frac{\frac{1}{2} - d}{E(1 - d)}\right]^{1/z}. \qquad (6)$$

The ESS is locally stable (i.e. the second partial derivative is less than one) when $2p(N+1)/p_i(x+1)$ is greater than one, which is whenever N is large.

(b) Graphical results of the model

A feel for the ESS given in equation (6) can be gained for the special case of selfing by examination of figure 5, where the equilibrium selfing rate (p^*) is plotted as a function of inbreeding depression (d) and the effects that parasites have on host fitness (E) for various exponents (z). For small exponents (e.g. $z = 0.1$), in which the transfer of infection increases very rapidly in mostly outcrossing population, there seems to be two alternative stable states, complete selfing and complete outcrossing, with a very sharp zone of transition between them (figure 5a). Complete selfing is favoured when inbreeding depression (d) and parasite effects (E) are both low (< 0.40). Complete outcrossing is favoured when the inbreeding depression is high (> 0.45) and when there are moderate to severe effects of parasites. Small exponents giving rise to rapid increases in the infection of selfed progeny in mostly outcrossed populations might be expected when there are

aggregated sibships, especially when there are few loci involved in disease resistance. Aggregated sibships would allow for the direct transfer of disease among closely related individuals as suggested by Rice (1983), Augspurger (1983), Shykoff & Schmid-Hempel (1991) and Herre (1993). Larger exponents ($z = 0.5-2.0$), however, can easily give rise to mixed (selfed and outcrossed) mating systems for a large fraction of the parameter space. For the present model, mixed reproduction is an unbeatable strategy for moderate to severe effects of parasites, provided inbreeding depression is less than 0.5. For moderate inbreeding depression ($d = 0.4-0.5$), mixed mating is stable even when there are small ($E = 0.1-0.3$) effects of parasites (figure 5b–d).

Thus the results of this simple model of allocation-dependent fitnesses suggest that mixtures of selfed and outcrossed progeny can be evolutionarily stable over a wide range of conditions, and may be a partial explanation for mixed mating in natural populations. Although the general result may hold, it is none the less clear that detailed genetic models will be helpful to determine (i) the effects of different modes of inbreeding depression (see Jarne & Charlesworth (1993) for a thorough review) and (ii) the effect of repeated selfing on inbreeding depression (see Campbell 1986; Lande & Schemske 1985; Charlesworth et al. 1990). Such models would also aid in evaluating the oversimplification presented here that cross-fertilized plants are relatively uninfected, which depending on the level of selfing and the genetics of infection may not be met. This is especially true if the advantages to outcrossing decline with increases in the population-wide selfing rate. A glimpse of this effect, however, can be seen by letting

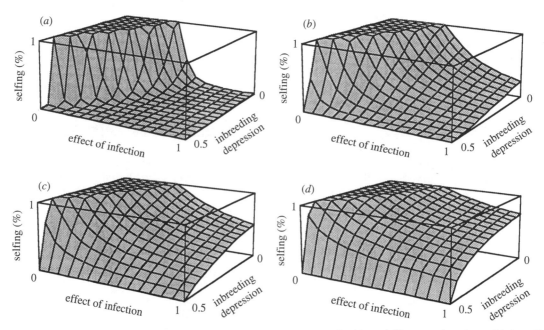

Figure 5. Graphical results for the evolutionarily stable percent of selfing (p^*) as a function of inbreeding depression (d) and the effects of parasites (E), as given in equation (6): (a) $z = 0.1$; (b) $z = 0.5$; (c) $z = 1.0$; (d) $z = 2.0$. Note that mixtures of selfed and outcrossed offspring are evolutionarily stable for large regions of the parameter space in (b–d).

the fitness gains through outcrossed progeny (W_m and W_f) vary with the mean selfing rate (p). This is easily accomplished by multiplying equations 2a and 2b by ($1 - ehp^x$), where h is the frequency of outcrossed progeny that are infected relative to uniparental progeny. Under these conditions, the equilibrium selfing rate in a large population becomes:

$$p^* \approx \left[\frac{1-2d}{E(2-2d-h)} \right]^{1/z}. \tag{7}$$

The results are shown graphically in figure 6 for h equal to one half (i.e. outcrossed progeny are half as infected as uniparental progeny). As expected, the region of parameter space increases for which complete selfing is evolutionarily stable, but mixed reproduction is none the less similarly stable over a significant portion of the same parameter space.

4. DISCUSSION

Parasites represent a potentially powerful source of selection on host reproductive strategies, especially if they can evolve rapidly enough to track the spread of common host genotypes. However, the parasite (or Red Queen) hypothesis for the maintenance of sex suffers from two general criticisms. One is that, to overcome the twofold advantage that clones may enjoy, the effects of parasites would seem to need to be severe if they are to keep the clone from fixing (May & Anderson 1983). One possible solution to this problem has been suggested by Hamilton *et al.* (1990). They argue convincingly that the effects of parasites can be exacerbated by intraspecific competition for resources. For example, infected plants in a greenhouse may show marginal reductions in fitness-related

traits when compared with control plants. However, when these same plants are grown under intense competition with uninfected plants, the effect of infection may be greatly increased owing to a synergism between competition and infection. In fact it seems reasonable to suspect that disease could be indirectly responsible for the failure of the most infected plants to reproduce under high levels of competition. Hamilton *et al.* (1990) show that this kind of synergism between competition and infection dampens the host–parasite gene-frequency oscillations, and gives sex an advantage over parthenogenesis. They refer to the synergism as rank-order truncation selection (see also the MIS model in Hamilton (1993)).

As an alternative, the accumulation of mutations through Muller's ratchet can also greatly reduce the severity of parasites needed to prevent the fixation of clones (Howard & Lively 1994). The basic idea rests on the assumption that clones start as rare mutants in large sexual populations; as such, they are initially very sensitive to mutation accumulation via Muller's ratchet. For mutation rates of one per genome per generation or greater, it is unlikely that a newly formed clone will survive for more than a few generations before its mutational load is significantly increased. The combination of mutation accumulation and moderate effects of debilitating parasites could prevent the fixation of clonal mutants in the short term. Oscillations in clonal frequency driven by parasites then lead to the further accumulation of mutations in clonal lineages and their eventual elimination.

It is worth noting that, under this kind of reasoning, any force that drives populations through periodic bottlenecks would have the same effect. Periodic

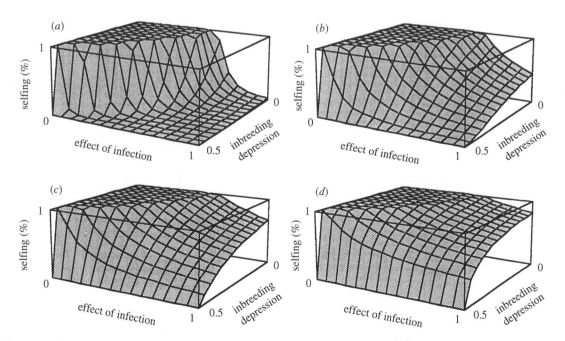

Figure 6. Graphical results for the evolutionarily stable percent of selfing (p^*) as a function of inbreeding depression (d) and the effects of parasites (E), as given in equation (7) for the relative frequency of infected sexual individuals (h) = 0.5: (a) $z = 0.1$; (b) $z = 0.5$; (c) $z = 1.0$; (d) $z = 2.0$. Compare with figure 5.

epidemics, for example, could favour sex, even if the disease does not have the specific genetic basis required by frequency-dependent selection. In fact population cycles driven by any source (e.g. physical disturbance, predation, or resource competition) would also favour cross-fertilization by driving clonal populations through ratchet-cranking bottlenecks. A major advantage (in our view) of parasite-mediated frequency-dependent selection is that it should predictably drive cycling and thereby reliably favour sex over parthenogenesis. This is true even if there are some refugia from infection, provided such refuges contain a sufficiently small fraction of the population so that the ratchet will work.

A second weakness of the parasite theory for sex is that parasites do not select for sex *per se*, but for rarity, however generated (figure 1). Hence, any theory of sex must be able to cope with the possibility of repeated mutations to parthenogenetic reproduction. We have found that the interaction between parasites and mutation accumulation can operate to render sex stable to invasion by multiple, independently derived clones (figures 2 and 3). The idea here is that, if the clone fails to fix, it begins to oscillate with the sexual population. However, during these oscillations, each time the clone is depressed to a low frequency by the parasite, it is subjected to the accumulation of further mutations through the action of the ratchet. This leads to the rapid extinction of clones (Howard & Lively 1994) and the stability of sexual reproduction when there is repeated mutation to parthenogenesis (figures 2 and 3).

As an aside, it is worth pointing out that rank-order truncation selection and parasite-aided mutation accumulation are not mutually exclusive alternatives. In fact, it would seem that both should operate in the wild. Moreover, it seems reasonable to think that they would interact in a synergistic way if competition for resources exacerbates the effects of infection, thereby slowing the growth rate of any clonal mutant and thus allowing for a greater rate of mutation accumulation. Another kind of synergism that merits future consideration is that between parasite load and mutational load. If, for example, individuals with high mutational loads were sicker when infected, then mutational load and parasites would have a synergistic, negative effect on host fitness.

It now seems that parasites should be at least part of the reason why dioecious outcrossing populations are stable to invasion by obligately parthenogenetic mutants. But what about hermaphroditic populations: can parasites prevent the fixation of alleles for self-fertilization? This is an important question for at least two reasons. One is that the 'cost of sex' is different. In a competition between obligately parthenogenetic and obligately sexual females, the cost of sex is proportional to the investment that sexual females put into sons (Maynard Smith 1971, 1978); relatedness is not relevant, except in how it might affect son production. However, for the self-fertilization of potentially outcrossing (non-cleistogamous) eggs, the cost of sex is due to the reduction in relatedness between the

parent and its outcrossed offspring (Williams 1975; Charlesworth 1980; Lloyd 1980; Lively & Lloyd 1990). The second, and more important, reason is that the evolution of self-fertilization in cosexual populations represents the indisputable evolution of uniparental reproduction at the individual level. Whereas obligate parthenogens and their sexual ancestors are reproductively isolated groups (Williams 1992), partly selfing individuals are not reproductively isolated from other partly selfing individuals. Any general theory of sex should be able to explain not only why sexual females are not replaced by obligately parthenogenetic females, but why cross-fertilization is not replaced by self-fertilization. In fact, the latter explanation would seem to be a more difficult challenge for the parasite theory of sex, because there is no single clone for the parasites to track in evolutionary time.

It may, of course, be that parasites are not needed (unless they are responsible for genetic load). Inbreeding depression seems to be universally associated with self-fertilization of individuals from outcrossing populations; if inbreeding depression reduces the fitness of selfed progeny by more than 50%, then inbreeding depression alone is sufficient for the maintenance of sex (Lloyd 1979). If, however, inbreeding depression reduces the fitness of selfed offspring by less than 50%, then selfing should have an advantage (Fisher 1941; Lloyd 1979). Hence, in the absence of additional factors, there are two alternative stable states that depend on inbreeding depression: complete selfing is stable if inbreeding depression is low, and complete outcrossing is stable if inbreeding depression is high. However, because mutation loads and inbreeding depression vary among individuals, complete selfing may be a stronger attractor (Lande & Schemske 1985).

The models presented here add parasites to inbreeding depression as a factor contributing to the 'cost' of selfing. They make the simplest possible assumption that the spread of parasites is related to the degree of self-fertilization in a population, which could result from either the reduction in heterozygosity or genetic diversity (or both) caused by selfing. The results depend on how the selfing rate maps onto the transfer of parasites. If small increases in selfing from an outcrossing ancestral state lead to dramatic increases in the transfer of disease (small values of z in equation (6)), then selfing and crossing seem to be alternative stable states as suggested by Lloyd (1979) and Lande & Schemske (1985). The main difference is that, with the addition of parasites, crossing is stable for values of inbreeding depression less than 50% (figure 5a). If, however, small increases in selfing lead to significant, but less dramatic, increases in the likelihood of infection, then mixtures of selfing and outcrossing are expected at the ESS for a large range of parameter values (figure 5b–d). This explanation is fundamentally different from the argument for time-lagged selection against common clones discussed above. Parasites might none the less explain mixed mating systems in plants and animals where inbreeding depression is less than 50%.

Finally, it seems reasonable to suggest that parasites could select for temporal mixtures of uniparental and biparental offspring, as observed in many facultatively parthenogenetic and selfing animals. Such selection would require that there are reliable cues indicative of the risk of parasite attack, such as high population density. In cyclically parthenogenetic zooplankton, for example, high population density is known to induce the production of males and sexual females (review in Bell 1982). Before such induction, genetic diversity in the population could be eroded by interclonal competition at the same time that population size is increasing. This is precisely the kind of situation that should favour the spread of disease and a facultative switch to cross-fertilization. Similarly, warmer water temperatures have been shown to induce a cross-fertilizing morph of an otherwise selfing hermaphroditic snail (Schrag & Read 1992; Schrag *et al.* 1994*b*). Such temperatures are indicative in some populations of an enhanced risk of attack by digenetic trematodes that castrate infected individuals (Schrag *et al.* 1994*a*).

We thank Lynda Delph, Mark Dybdahl, Steve Johnson and Andy Peters for helpful comments on the manuscript. We are especially grateful to Bill Hamilton for detailed comments on the computer simulations, and to Phillipe Jarne for constructive insights on the selfing model. David Lloyd and Lynda Delph are also gratefully acknowledged for many patient hours of discussion on breeding-system evolution. This study was supported by a U.S. National Science Foundation grant (BSR-9008848) to C.M.L.

REFERENCES

Augspurger, C.K. 1983 Seed dispersal of the tropical tree, *Platypodium elegans*, and the escape of its seedlings from fungal pathogens. *J. Ecol.* **71**, 759–771.

Bell, G. 1982 *The masterpiece of nature: the evolution and genetics of sexuality.* Berkeley, California: University of California Press.

Bell, G. 1988 *Sex and death in the protozoa.* Cambridge University Press.

Bell, G. & Maynard Smith, J. 1987 Short-term selection for recombination among mutually antagonistic species. *Nature, Lond.* **328**, 66–68.

Bierzychudek, P. 1987 Patterns in plant parthenogenesis. In *The evolution of sex and its consequences* (ed. S. C. Stearns), pp. 197–218. Basle: Birkhäuser Verlag.

Bremermann, H.J. 1980 Sex and polymorphism as strategies in host-pathogen interactions. *J. theor. Biol.* **87**, 641–702.

Campbell, R.B. 1986 The interdependence of mating structure and inbreeding depression. *Theor. Popul. Biol.* **30**, 232–244.

Chao, L. 1990 Fitness of RNA virus decreased by Muller's ratchet. *Nature, Lond.* **348**, 454–455.

Charlesworth, B. 1980 The cost of sex in relation to mating system. *J. theor. Biol.* **84**, 655–671.

Charlesworth, D., Morgan, M.T. & Charlesworth, B. 1990 Inbreeding depression, genetic load, and the evolution of outcrossing rates in a multilocus system with no linkage. *Evolution* **44**, 1469–1489.

Fisher, R.A. 1941 Average excess and average effect of a gene substitution. *Ann. Eugenics* **11**, 53–63.

Glesener, R.R. & Tilman, D. 1978 Sexuality and the components of environmental uncertainty: Clues from geographic parthenogenesis in terrestrial animals. *Am. Nat.* **112**, 659–673.

Hamilton, W.D. 1967 Extraordinary sex ratios. *Science, Wash.* **156**, 477–488.

Hamilton, W.D. 1980 Sex versus non-sex versus parasite. *Oikos* **35**, 282–290.

Hamilton, W.D. 1982 Pathogens as causes of genetic diversity in their host populations. In *Population biology of infectious diseases* (ed. R. M. Anderson & R.M. May), pp. 269–296. New York: Springer-Verlag.

Hamilton, W. D. 1993 Haploid dynamic polymorphism in a host with matching parasites: effects of mutation/subdivision, linkage, and patterns of selection. *J. Hered.* **84**, 328–338.

Hamilton, W. D., Axelrod, R. & Tanese, R. 1990 Sexual reproduction as an adaptation to resist parasites (a review). *Proc. Natn. Acad. Sci. U.S.A.* **87**, 3566–3573.

Herre, E.A. 1993 Population structure and the evolution of virulence in nematode parasites of fig wasps. *Science, Wash.* **259**, 1442–1445.

Holsinger, K., Feldman, M.W. & Christiansen, F.B. 1984 The evolution of self fertilization in plants: A population genetic model. *Am. Nat.* **124**, 446–453.

Houle, D., Hoffmaster, D.K., Assimacopoulos, S. & Charlesworth, B. 1992 The genomic mutation rate for fitness in *Drosophila. Nature, Lond.* **359**, 58–60.

Howard, R.S. & Lively, C.M. 1994 Parasitism, mutation accumulation and the maintenance of sex. *Nature, Lond.* **367**, 554–557. (Reprinted figures in vol. **368**, p. 358.)

Hutson, V. & Law, R. 1981 Evolution of recombination in populations experiencing frequency-dependent selection with time delay. *Proc. R. Soc. Lond.* B **213**, 345–359.

Jaenike, J. 1978 An hypothesis to account for the maintenance of sex within populations. *Evol. Theory* **3**, 191–194.

Jarne, P. & Charlesworth, D. 1993 The evolution of the selfing rate in functionally hermaphrodite plants and animals. *A. Rev. Ecol. Syst.* **24**, 441–466.

Lande, R. & Schemske, D.W. 1985 The evolution of self-fertilization and inbreeding depression in plants. I. Genetic models. *Evolution* **39**, 24–40.

Leslie, J.F. & Vrijenhoek, R.C. 1980 Consideration of Muller's ratchet mechanism through studies of genetic linkage and genomic compatibilities in clonally reproducing *Poeciliopsis. Evolution* **34**, 1105–1115.

Levin, D. 1975 Pest pressure and recombination systems in plants. *Am. Nat.* **109**, 437–451.

Lively, C.M. 1987 Evidence from a New Zealand snail for the maintenance of sex by parasitism. *Nature, Lond.* **328**, 519–521.

Lively, C.M. 1992 Parthenogenesis in a freshwater snail: reproductive assurance versus parasitic release. *Evolution* **46**, 907–913.

Lively, C.M. & Lloyd, D.G. 1990 The cost of biparental sex under individual selection. *Am. Nat.* **135**, 489–500.

Lloyd, D.G. 1979 Some reproductive factors affecting the selection of self-fertilization in plants. *Am. Nat.* **13**, 67–79.

Lloyd, D.G. 1980 Benefits and handicaps of sexual reproduction. *Evol. Biol.* **13**, 69–111.

Lynch, M. & Gabriel, W. 1990 Mutational load and the survival of small populations. *Evolution* **44**, 1725–1737.

May, R.M. & Anderson, R.M. 1983 Epidemiology and genetics in the coevolution of parasites and hosts. *Proc. R. Soc. Lond.* B **219**, 281–313.

Maynard Smith, J. 1971 The origin and maintenance of sex. In *Group Selection* (ed. G. C. Williams), pp. 163–175. Chicago, Illinois: Aldine Atherton.

Maynard Smith, J. 1978 *The evolution of sex.* Cambridge University Press.

Maynard Smith, J. 1982 *Evolution and the theory of games.* Cambridge University Press.

Mukai, T., Chigusa, S.T., Mettler, L.E. & Crow, J.F. 1972 Mutation rate and dominance of genes affecting viability in *Drosophila melanogaster. Genetics, Princeton* **72**, 335–355.

Muller, H.J. 1964 The relation of recombination to mutational advance. *Mutat. Res.* **1**, 2–9.

Nee, S. 1989 Antagonistic coevolution and the evolution of genotypic randomization. *J. theor. Biol.* **140**, 499–518.

Rice, W.R. 1983 Parent-offspring transmission: a selective agent promoting sexual reproduction. *Am. Nat.* **121**, 187–203.

Rice, W.R. 1994 Degradation of a nonrecombining chromosome. *Science, Wash.* **263**, 230–232.

Shykoff, J.A. & Schmid-Hempel, P. 1991 Parasites and the advantage of genetic variability within social insect colonies. *Proc. R. Soc. Lond.* B **243**, 55–58.

Schrag, S.J., Mooers, A.O., Ndifon, G.T. & Read, A.F. 1994*a* Ecological correlates of male outcrossing ability in a simultaneous hermaphordite snail. *Am. Nat.* **143**, 636–655.

Schrag, S.J., Ndifon, G.T. & Read, A.F. 1994*b* Temperature determination of male outcrossing ability in wild populations of a simultaneous hermaphrodite snail. *Ecology.* (In the press.)

Schrag, S.J. & Read, A.F. 1992 Temperature determination of male outcrossing ability in a simultaneous hermaphrodite gastropod. *Evolution* **46**, 1698–1707.

Weismann, A. 1989 *Essays upon heredity and kindred biological problems* (transl. E. B. Poulton, S. Schonland & E. E. Shipley). Oxford: Clarendon Press.

Williams, G.C. 1975 *Sex and evolution.* Princeton, New Jersey: Princeton University Press.

Williams, G.C. 1992 *Natural selection: domains, levels, and challenges.* Oxford University Press.

Discussion

J. GODFREY (*41 Lawford Road, London, U.K.*). Dr Lively's argument that the mutation rate has a crucial influence on the relative advantage of sexual and asexual reproduction needs to take account of the probable effect of latitude on mutation. In the laboratory mutation goes up with an increase in temperature, exhibiting a Q_{10} like other chemical processes. So natural populations near to the equator could be expected to have a higher mutation rate than related populations nearer to a pole. This effect has not been investigated as far as I know. If natural selection adjusts the mutation rate towards an optimum, the effect would be less than a doubling of the mutation rate for each 10°C increase in temperature. To test whether this is so would require stoic study of the mutation rate of natural populations of species with wide distribution. If there is a latitudinal effect, could it explain the fact that a majority of carbon has been accumulated by asexual plants in parts of the world far from the tropics? May the Red Queen not race in a circle, but rather in a spiral, trending towards the equator?

C. M. LIVELY. This is an extraordinary suggestion. If mutation rates truly are greater at lower latitudes, then this could help explain, under our model, why sex is more common in these regions. It would also greatly aid Kondrashov's models for the evolution of sex by giving some explanation for the ecological distribution of cross-fertilizing species. I wholly agree that investigation into the possibility of such an effect would be worthwhile.

P. HIGGS (*University of Sheffield, U.K.*). Dr Lively discussed modelling of both Muller's ratchet, and host–parasite coevolution, and appeared to conclude that a combination of both processes was necessary to give sexual reproduction an advantage over parthenogenesis. Muller's ratchet on its own has sometimes been proposed as a reason for the advantage of sex and recombination. Are there any situations in his model where the ratchet alone is sufficient? In his simulations, increasing the mutation rate gave an advantage to sex. Is there a simple reason why this should be the case? Muller's ratchet and host–parasite coevolution both share the common feature that the population is always evolving, and gene frequencies never reach a mutation selection balance. This is in contrast to the theory that the reason for the advantage of sex is due to the higher equilibrium fitness of the sexual population in certain fitness landscapes with epistatic interactions once mutation-selection balance has been reached. It would seem to be an important feature of real evolutionary processes that gene frequencies never have chance to reach equilibrium.

C. M. LIVELY. There were no situations in our simulations where Muller's ratchet was sufficient by itself to confer evolutionary stability on a sexual population. For the mutation rates and selection coefficients we used (which seem reasonable given what is presently known), the ratchet simply worked too slowly to prevent a clone with a twofold reproductive advantage from replacing the descendants of its sexual ancestor. The explanation is that the ratchet-like mechanism described by Muller slows down after the clonal lineage passes through its initial phase of growth from a single individual to several hundred individuals. The reason that parasites are so effective in driving the ratchet is that they periodically and predictably depress the size of the clonal population, and this increase the rate at which the ratchet 'clicks'. Hence, each time the parasites drive the clone through a population-size bottleneck, it emerges with a greater load and an eroded ability to compete with the sexual population. Eventually the clone cannot replace itself, and it goes extinct.

It is certainly true, as suggested, that the ratchet allows a clonal population to accumulate more mutations than they would otherwise have at mutation selection balance, while the sexual population has much less difficulty in this regard. It is also true that host–parasite interactions lead to dynamic cycles rather than static equilibria, and that understanding these cycles is the key to understanding the Red Queen theory for sex.

J. SHYKOFF (*Swiss Federal Institute of Technology, Zurich, Switzerland*). Using the model for the effects of parasites on the evolution of partial selfing, can Dr Lively distinguish between an individual mixed ESS and a population mixed ESS?

Individual mixed ESS versus population mixed ESS will explain different kinds of breeding systems. Some plant breeding systems such as gynodioecy represent a population mixed strategy where each individual follows a pure strategy. Other breeding systems such as gynomonoecy in plants or aphally in snails represent individual mixed strategies. Which types of breeding system will this model help to explain?

C. M. LIVELY. That is an interesting question. Dr Shykoff is asking whether the population could be dimorphic, so that some individuals are selfing, whereas others are outcrossing; or, in more formal terms whether the population is at a true evolutionary stable strategy or at an evolutionarily stable state. In our models, we performed the standard second derivative test, which indicated that the mixture of selfed and outcrossed offspring would occur within individuals, meeting the condition for a true ESS.

J. D. GILLETT (*London School of Hygiene and Tropical Medicine, U.K.*). I would like to ask three questions. First. Why confine this discussion to the threat from small enemies only? Surely, large or small, both make a living by exploiting the tissues of the host, one way or another: *Panthera leo*, for example, exploiting mainly muscle, *Plasmodium falciparum*, haemoglobin. Sex provides the plasticity – the polymorphism – to cope with threats of all dimensions and kinds, large or small. The means may differ, but the result is the same: survival by at least some of those attacked. A greenfly puts up one kind of defence against a virus and another against a ladybird; each is suited to the kind of threat and each is inherited.

Second. If parasitic disease, including those of viral origin, were the main threat that sex guards against then why do so many hundreds of species of insects in the temperate regions cease sexual reproduction, replacing it with asexual reproduction or parthenogenesis, during the summer months just when these threats are surely at their highest?

Third. What about bacteria, subject to attack by very many different viruses; how do they get by without sexual reproduction? Their plasticity of response must be through other channels.

C. M. LIVELY. Regarding Dr Gillett's first question, I did not mean to imply that the size of the enemies matters.

What seems most important is that the enemy attacks in a frequency-dependent matter, and that the attack is lagged in time. It is the time lags that set up the dynamic oscillations that are so important to the Red Queen theory. It would seem to me, however, that small enemies (pathogens and parasites) are more likely than predators to attack in this way. This does not mean that predation is not sometimes dependent on the frequencies of prey genotypes; it would just seem to be much less common, especially for a predator like *Panthera leo*.

Your second question is a very good one. One of the biggest challenges for any theory of sex is to explain cyclical parthenogenesis. Under the parasite theory of sex, one would expect that a switch to sex should occur when the risk of disease is greatest. In freshwater zooplankton, this switch often occurs when population densities become high and resources are limited. This result could be used to support models for genetic polymorphism and sex which are based on resource competition in heterogeneous environments. But it is also possible that the risk of disease increases with host density, and that host density is used as a cue for the future risk of parasitic attack. Hence, although there is much to learn about the kinds of parthenogenesis mentioned, it would seem premature at present to reject the parasite theory.

Regarding sex in bacteria, I don't think they should be able to get away without some kind of genetic exchange if they are threatened by persistent attack by viruses, unless, of course, their mutation rates to alternative defensive types are very common.

2

Recognition and polymorphism in host–parasite genetics

STEVEN A. FRANK

Department of Ecology and Evolutionary Biology, University of California, Irvine, California 92717, U.S.A.

SUMMARY

Genetic specificity occurs in many host–parasite systems. Each host can recognize and resist only a subset of parasites; each parasite can grow only on particular hosts. Biochemical recognition systems determine which matching host and parasite genotypes result in resistance or disease. Recognition systems are often associated with widespread genetic polymorphism in the host and parasite populations. I describe four systems with matching host–parasite polymorphisms: plant–pathogen interactions, nuclear–cytoplasmic conflict in plants, restriction enzymes in bacterial defence against viruses, and bacterial plasmids that compete by toxin production and toxin immunity. These systems highlight several inductive problems. For example, the observed patterns of resistance and susceptibility between samples of hosts and parasites are often used to study polymorphism. The detectable polymorphism by this method may be a poor guide to the actual polymorphism and to the underlying biochemistry of host–parasite recognition. The problem of using detectable polymorphism to infer the true nature of recognition and polymorphism is exacerbated by non-equilibrium fluctuations in allele frequencies that commonly occur in host–parasite systems. Another problem is that different matching systems may lead either to low frequencies of host resistance and common parasites, or to common resistance and rare parasites. Thus low levels of host resistance or rare parasites do not imply that parasitism is an unimportant evolutionary force on host diversity. Knowledge of biochemical recognition systems and dynamical analysis of models provide a framework for analysing the widespread polymorphisms in host–parasite genetics.

1. INTRODUCTION

Recognition systems often have spectacular genetic polymorphisms. Self-incompatibility or mating-type loci may have 100 or more alleles, each with a distinct label (Fincham *et al.* 1979; Richards 1986). The great diversity of major histocompatibility complex (MHC) alleles is probably maintained by the need to recognize a wide range of parasitic invaders (Potts & Wakeland 1993). The numerous bacterial restriction enzymes that revolutionized molecular biology evolved for specific recognition and defence against parasitic attack (Wilson & Murray 1991).

In this paper, I show that the study of recognition and polymorphism in host–parasite systems requires analysis of five related factors, as follows.

1. The biochemistry of recognition: the interacting molecules of the host and parasite that signal a parasitic invasion to the host's defensive arsenal.
2. The potential genetic diversity: the number of host and parasite genotypes that have different recognition properties.
3. The pattern of specificity: which parasite genotypes can attack which host genotypes.
4. The dynamics of disease: the processes that combine the recognition system and ecology to determine the actual amount of genetic polymorphism at any point in time and space.
5. Inductive problems: the pitfalls of using observed polymorphism and specificity to infer factors 1–4.

I focus on four exemplar systems of recognition and polymorphism: plant–pathogen genetics, nuclear–cytoplasmic conflict in plants, bacterial restriction enzymes that defend against viruses, and bacterial plasmids that compete by toxin production and toxin immunity. In each of these systems the hosts (those attacked) are widely polymorphic for their ability to recognize and resist assault. The parasites (aggressors) are, in turn, widely polymorphic for host-range: the range of host genotypes that they can attack by escaping recognition and resistance. These four examples are perhaps the best understood of systems with reciprocal interactions between genetic polymorphisms of host and parasite. 'Best understood' is, of course, a relative measure. For these examples one can describe a plausible system of recognition and reciprocal interaction to explain the observed polymorphisms, but many details are unknown.

I have two goals. First, I summarize the natural history of these four systems. The facts are interesting, and the examples provide a database in which to search for recurring themes in host–parasite genetics.

(a)

(b)

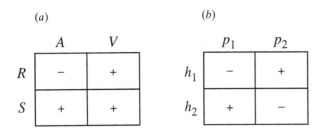

Figure 1. Resistance (−) and susceptibility (+) between two host alleles and two pathogen alleles. (a) Gene-for-gene specificity. (b) Matching-allele specificity.

Second, I focus on inductive problems and show that, in host–parasite systems, appearances are often deceptive. I provide examples for the following problems, each of which has led investigators to draw incorrect conclusions from their data.

1. Detectable polymorphism is a poor guide to the actual polymorphism.
2. Detectable polymorphism is a poor guide to the underlying biochemistry of recognition.
3. A low frequency of host resistance and a wide parasite host-range do not imply a lack of coevolutionary pressure.
4. A low level of parasitism does not imply a lack of parasite pressure on host polymorphism.

Dynamic models for each of the individual examples have been published elsewhere; I do not repeat the details here. Instead I emphasize general aspects of host–parasite specificity and evolutionary dynamics that contribute to misleading inferences about recognition and polymorphism.

2. PLANTS AND PATHOGENS

The reciprocal genetic interactions between plants and pathogens have been studied extensively because of the economic importance of crop disease. Recent work has extended the study of plant disease to natural populations. I will focus on two patterns of plant–pathogen genetics in natural populations (Burdon 1987; Frank 1992). (1) Genetic specificity is common. Each host genotype resists only specific pathogen genotypes; each pathogen genotype attacks only specific host genotypes. Many host and pathogen loci interact to determine the success of a pathogen attack. (2) Each host locus has a low frequency of resistance alleles, and each matching pathogen locus has a high frequency of alleles with a broad host-range.

I will show that two distinct models of specificity, gene-for-gene and matching-allele, are consistent with these observations on recognition and polymorphism. These alternative models illustrate the inductive problems of using detectable polymorphisms to infer the actual polymorphism and the biochemistry of specificity.

(a) Gene-for-gene models

During the 1940s and 1950s, H. H. Flor studied the inheritance of specific resistance and virulence factors

in flax and its fungal pathogen, flax rust (Flor 1971). The interaction between host and pathogen genotypes turned out to have simple properties that Flor referred to as a 'gene-for-gene' system. In an idealized gene-for-gene system, each pair of resistance and susceptibility alleles in the host has a matching pair of host-range alleles in the pathogen.

Figure 1a shows the standard gene-for-gene model (Flor 1956, 1971; Burdon 1987). The host has two phenotypes, resistant (R) and susceptible (S). The pathogen has two phenotypes, avirulent (A) and virulent (V). Plant pathologists use the term 'virulence' for host-range. I follow their convention.

In the gene-for-gene model a host resists attack only when the host–parasite pair has an R : A match. Person & Mayo (1974) refer to this match as a 'stop-signal'. Recent biochemical models suggest that the avirulence allele (A) produces a gene product (elicitor) that can be recognized only by specific host receptors (R). This specific elicitor–receptor recognition induces a non-specific set of host defence mechanisms (Gabriel & Rolfe 1990).

In multilocus interactions each host locus is matched to a single parasite locus. If there is an R : A match between at least one pair of host–parasite loci, then disease does not develop. This is consistent with the model of the R : A match as a stop-signal.

Although the relationship between specific factors is simple in a gene-for-gene system, the total interaction between a host and its pathogen is complex. Flor and others have identified 29 separate host resistance factors in flax, each with a complementary virulence factor in flax rust (Flor 1971; Lawrence et al. 1981). Similar gene-for-gene interactions are now known or suspected for over 25 different host–pathogen pairs (Burdon 1987). These systems do not conform exactly to the idealized gene-for-gene assumptions (Christ et al. 1987), but these systems do have complementary major-gene interactions between hosts and pathogens.

Several theoretical models have been developed to explain observed polymorphism in systems that are believed to have gene-for-gene specificity (reviewed by Leonard & Czochor 1980; Levin 1983; Burdon 1987; Frank 1992, 1993a). These models assume that virulence alleles have a negative effect on fitness that offsets the benefit of wider host-range. This assumption is necessary because, without a fitness cost, the virulence allele would spread to fixation (in figure 1a, V has an advantage over A). A model that predicted fixation of virulence could not explain the observed polymorphism of virulence and avirulence alleles (Vanderplank 1968). A similar argument leads to the conclusion that resistance must have a cost.

Models with costs predict that resistance alleles will be rare relative to susceptibility alleles, and virulence alleles will be common relative to avirulence alleles. Observations appear to support the predicted frequencies of these allelic classes (reviewed by Frank 1992, 1993a). These observations are not surprising given the assumptions of the gene-for-gene models. A pathogen must avoid recognition at all pairs of matching host–parasite loci in order to succeed. This requires few recognizable avirulence alleles and many costly universal host-range (virulence) alleles. For a host the costs of resistance alleles must be balanced against the low probability of matching rare avirulence alleles. This balance usually leads to a low frequency of the resistance allele at each locus.

Small costs of resistance and virulence lead at equilibrium to widespread multilocus diversity, pathogens that can attack a wide range of hosts, and consequently little effective resistance. In spite of the low level of resistance, the widespread genetic diversity is maintained by reciprocal coevolutionary pressures of host and parasite.

The gene-for-gene model is compelling because it is consistent with several observations. First, the bio-chemistry of recognition, the elicitor–receptor model, fits the gene-for-gene model in figure 1*a* if universal virulence (V) is the absence of an elicitor and universal susceptibility (S) is the absence of a matching receptor.

Second, the gene-for-gene model, which specifies the types of alleles and their phenotypic effects, is consistent with the observed pattern of phenotypic interactions shown in figure 1*a*. This reasoning may appear circular because the gene-for-gene model was derived from the observed patterns of phenotypic interactions. However, as I show in the next section, a very different model for the genetics of host–parasite interactions would also lead to a similar pattern of phenotypic interactions.

Third, the population genetic observations of rare resistance alleles and common virulence alleles can be explained by gene-for-gene interactions plus small costs of resistance and virulence. There is only limited evidence for such costs, but small and essentially undetectable costs would be sufficient to explain the observed polymorphisms.

(b) *Matching-allele models*

The gene-for-gene model is convincing because of the range of observations explained. However, a simple alternative, the matching-allele model, can also explain the same observations (Frank 1993*b*). In a matching-allele model with a single haploid locus, each of the n host alleles causes recognition and resistance to only one of the n parasite alleles (Frank 1991*a*). Thus each host is resistant to $1/n$ of the parasite genotypes, and each parasite can attack $(n-1)/n$ of the host genotypes. Figure 1*b* illustrates the matching-allele model for $n = 2$.

In the matching-allele model each parasite genotype functions as either an avirulence allele or a virulence allele depending on the host genotype. By contrast, the gene-for-gene system always has a universal virulence allele that can attack all host genotypes.

Similarly, each host genotype in the matching-allele model functions as either a resistance or a suscept-ibility allele depending on the parasite genotype. The classical gene-for-gene system always has a universal susceptible genotype that can be attacked by all parasite genotypes.

In terms of biochemical recognition, matching alleles assume a one-to-one correspondence between elicitors and receptors. In the gene-for-gene model the universal virulence allele does not produce an elicitor that can be recognized by any of the available host receptors. Similarly, the universal susceptibility allele does not produce a receptor that can recognize any of the available pathogen elicitors.

Both the matching-allele and gene-for-gene assumptions are consistent with the elicitor–receptor model for the biochemistry of recognition. Matching-allele and gene-for-gene models differ in the range of elicitors and receptors that may exist at each matching pair of host–parasite loci. Thus bio-chemical evidence on recognition does not distinguish between the models. However, the phenotypic observations of susceptibility and resistance follow the gene-for-gene model (figure 1*a*) rather than the matching-allele model with $n = 2$ (figure 1*b*). This is not surprising because the gene-for-gene model was derived from the phenotypic pattern in figure 1*a*. Appearances can be deceptive, however. One must consider what type of genetic system and pattern of polymorphism would be inferred from samples of the host and parasite populations. The standard procedure is to isolate some host and parasite lines and then test each host against each parasite for resistance or susceptibility.

Here is a reasonable method of classification (see figure 2). (i) Find the host genotype that resists the highest proportion of parasites in the sample. Label that host genotype R for resistant. Only those hosts that resist exactly the same set of parasites are classified as R. (ii) Label all other hosts as S for susceptible. (iii) Label all parasites that cannot attack host genotype R as A for avirulent. (iv) Label all other parasites as V for virulent.

Figure 2 shows that, after following this procedure, one has a classification that is similar to the gene-for-gene system in figure 1a. In figure 2, the h_1 allele is classified as R and the matching p_1 allele is classified as A. What is the frequency of host–parasite pairs that would be misclassified by a gene-for-gene model if the

		A	V		
		p_1	p_2	p_3	p_4
R	h_1	−	+	+	+
S	h_2	+	−	+	+
	h_3	+	+	−	+
	h_4	+	+	+	−

Figure 2. Resistance (−) and susceptibility (+) in a matching allele model with four alleles. It is assumed that the $h_1 : p_1$ match has a frequency greater than or equal to any other diagonal match. Thus $h_1 : p_1$ is arbitrarily labelled as the $R : A$ match for the procedure outlined in the text.

true system were a matching-allele model with n alleles? The R and A alleles were defined strictly by their response in the sample, so there can be no errors in any host–parasite pair in which the host is classified as R or the parasite as A. All errors must occur when a host–parasite pair, classified as $S : V$, yields a resistant reaction rather than the predicted susceptible response. These errors occur only on the diagonal elements of the $(n-1) \times (n-1)$ submatrix of $S : V$ in figure 2. The frequency of errors in the entire table is $(n-1)/n^2$ at equilibrium. Surprisingly, the data fit the gene-for-gene pattern more closely as the number of alleles, n, increases. (Details and further discussion are in Frank (1993*b*).)

The matching-allele model is consistent with the elicitor–receptor model of biochemical specificity and with the gene-for-gene pattern of phenotypic specificity. The next issue concerns the observed patterns of rare resistance alleles, common virulence alleles, and a low phenotypic frequency of resistance. In the matching-allele model each host allele resists only $1/n$ of the alternative pathogen alleles. The observed frequency of host alleles classified as resistance (R) will tend to be low, and the observed frequency of pathogen alleles classified as virulence (V) will tend to be high (figure 2). Thus the matching-allele model can explain the observed polymorphism and frequency of resistance without invoking costs of resistance and virulence.

These models illustrate three problems (Frank 1993*b*). First, detectable polymorphism is a poor guide to the actual polymorphism. Second, detectable polymorphism is a poor guide to the underlying biochemistry of recognition. For example, if p_3 and p_4 were absent from a sample (figure 2), then host alleles h_3 and h_4 would be grouped as a single allelic type that is susceptible to all pathogen genotypes. This conclusion is misleading about the actual polymorphism and about the specificity of recognition. The third problem is that a low frequency of host resistance and a wide parasite host-range do not imply a lack of coevolutionary pressure. In both gene-for-gene and matching-allele models the frequency of resistance tends to be low even though the widespread polymorphism is maintained by coevolutionary interactions between host and parasite.

These inductive problems are not surprising, particularly after they have been illustrated with simple examples. None the less, they are common mistakes. The next section provides further evidence that detectable polymorphism is misleading when used to infer actual polymorphism and the biochemistry of recognition.

3. CYTOPLASMIC MALE STERILITY IN PLANTS

Most organisms inherit mitochondrial DNA from their mothers, with no input from their fathers. By contrast, most other genetic material is obtained equally from the mother and father. Typically these different modes of transmission, matrilineal versus biparental, have no consequences for the direction of evolutionary change favoured by selection. For example, efficient respiration increases both matrilineal and biparental transmission.

The allocation of resources to sons and daughters affects matrilineal and biparental transmission differently. Traits that enhance the production of daughters at the expense of sons always increase the transmission of matrilineally inherited genes. For example, in some hermaphroditic plants the mitochondrial genes may inhibit pollen development and simultaneously enhance the production of seeds (Edwardson 1970; Hanson 1991). Selection of genetic variants in the mitochondria would favour complete loss of pollen production in exchange for a small increase in seed production because the mitochondrial genes are transmitted only through seeds (Lewis 1941).

Reallocation of resources from pollen to seeds can greatly reduce the transmission of nuclear genes because biparental transmission depends on the sum of the success through seeds and pollen. Thus there is a conflict of interest between the mitochondrial (cytoplasmic) and nuclear genes over the allocation of resources to male (pollen) and female (ovule) reproduction (Gouyon & Couvet 1985; Frank 1989). Consistent with this idea of conflict, nuclear genes often restore male fertility by overcoming the male-sterility effects of the cytoplasm.

The nuclear–cytoplasmic conflict is very similar to a host–parasite system: there is antagonism over resources for reproduction, cytoplasmic (parasite) genes determine the host-range for exploitation, and cytoplasmic genes interact with nuclear (host) resistance genes to determine the specificity of the interaction. Cytoplasmic inheritance influences the patterns of 'parasite' transmission but, on the whole, the genetics and population dynamics are typical of host–parasite interactions (Gouyon & Couvet 1985; Frank 1989; Gouyon *et al.* 1991).

The reduction of pollen caused by cytoplasmic genes is called cytoplasmic male sterility (cms). Laser & Lersten (1972) list reports of cms in 140 species from 47 genera across 20 families. More than half of these cases occurred naturally, about 20% were uncovered by intraspecific crosses, and the rest were observed in interspecific crosses. Moreover, this listing is an underestimate of the true extent of cms because detecting a cytoplasmic component to a male sterile phenotype requires genetic analysis of polymorphism (see below).

Wild populations of cms maintain several distinct cytoplasmic genotypes (cytotypes). Each cytotype is capable of causing male sterility by an apparently different mechanism because each is susceptible to a particular subset of nuclear restorer alleles. Nuclear restorer alleles are typically polymorphic at several loci, with each allele specialized for restoring pollen fertility when associated with particular cytotypes. The observations are summarized in Frank (1989), Couvet *et al.* (1990) and Koelewijn (1993).

cms has reciprocal genetic specificity of nucleus and cytoplasm and widespread polymorphism. The basic questions of recognition and polymorphism are similar to those in other host–parasite systems. What is the relationship between the polymorphism that is detectable in a sample of plants and the actual distribution of genetic variation? How does the detectable number of genotypes compare with the potential range of polymorphic alternatives? What can be learned about the biochemistry of specificity from the detectable polymorphism?

The genetics of CMS have been studied by segregation ratios from crosses. These data are used to infer the number of cytotypes, the number of nuclear loci, the specificity of nuclear–cytoplasmic interactions in determining phenotype, and the amount of polymorphism in the sample. Without biochemical evidence of specificity, there is no other way to begin an analysis. However, the work is tedious and the segregation analysis is more intuition than algorithm. A final conclusion that, for example, there are at least three but perhaps many more nuclear loci involved provides little information beyond the fact that the multilocus genetics of specificity are complex.

The problem of inferring specificity occurs even if the underlying biochemistry is simple. To study this problem, I built a computer simulation with coevolving nucleus and cytoplasm. I set the number of loci and the specificity of the interaction. I studied a variety of specificities based on the stop-signal theories of plant–pathogen genetics. The idea is that a cytoplasmic male sterility allele interferes with a step in the pathway of pollen development. A matching nuclear allele restores pollen fertility by preventing the action of a specific cytoplasmic allele. (See Frank (1989) for detailed assumptions and a summary of the relevant biological observations.)

In the model I studied most intensively, I assumed that each cytoplasm has two loci. At each of these loci there are four alternative alleles, which are coded as the set $\{000, 001, 010, 100\}$. The haploid cytoplasmic genotype for the two loci is written by concatenating the pair of alleles, yielding a string with six bits, for example 010100. Each cytoplasmic 1 represents a different way in which the cytoplasm can interfere with pollen development. There are six matching nuclear loci. At each diploid locus the 1 allele can act as a restorer and the 0 allele has no effect, with 1 dominant to 0. The phenotype for the six loci can be written as a string with six bits, for example 101111. Each restorer locus is specific for one of the six possible methods of cytoplasmic interference in pollen production. A plant is male sterile if the cytoplasm has a single specificity not matched by a restorer. In terms of the bit-strings, if the cytoplasm has one or more unmatched 1s, then the plant is male sterile. The interaction between the example cytoplasmic and male sterility genotypes would yield male sterility because the 1 at the second (from the left) cytoplasmic locus is unmatched by its nuclear restorer locus.

Figure 3 illustrates some problems of inferring polymorphism and specificity. The columns in the left box show three different cytotypes (the cytoplasmic 'loci' do not recombine). The rows show the four nuclear phenotypes that occur when there is

polymorphism at the first and sixth loci. Perfectly executed crossing experiments and analysis yield the classification of male sterile (female) and male fertile (hermaphrodite) phenotypes in the figure, and the following conclusions. (i) There appear to be only two cytotypes because nuclear polymorphism in the sample cannot separate cytotypes B and C. Additional samples with more nuclear polymorphism would increase the inferred number of cytotypes. (ii) The first nuclear restorer locus appears to act independently of cytotype. This locus would be classified as a purely nuclear male sterility factor until a cytotype with a 0 in the first position was discovered. (iii) The system would be classified as purely nuclear control if the cytotype A were absent.

The right-hand box in figure 3 emphasizes a different set of problems. (i) Cytotype D appears to be a fully male-fertile cytotype. Additional nuclear polymorphism will change this inference about cytoplasmic effects. (ii) It appears that cytotype A requires simultaneous effects from two nuclear loci to restore pollen fertility. Additional cytoplasmic polymorphism would show that the nuclear loci have independent effects.

These examples and the further details in Frank (1989) highlight three problems of induction. First, the detectable polymorphism in a sample underestimates the actual polymorphism in the sample. Second, the inferred genetic system based on detectable polymorphism is a poor guide to the true specificity. Third, even the actual polymorphism in a sample is a poor guide to the number of loci and the specificity of the interaction. This occurs because nonequilibrium fluctuations in allele frequencies cause many alleles to be rare or absent at any particular point in time and space.

Why does it matter if the detectable and actual polymorphism differ in a sample? There is no problem if the goal is pattern description: a locus with no effect in the current context can be ignored when describing the current pattern. But suppose the goal is to explain why polymorphism is maintained and why the frequencies of genotypes and male sterile plants vary widely across space. The chain of reasoning is: detectable polymorphism → inferred specificity → model to explain evolutionary dynamics → predictions about polymorphism and spatial variation → comparison with detectable polymorph-

cytotypes

		A 100001	B 100010	C 100100		A 100001	D 010010
nucleus	011110	F	F	F		F	H
	011111	F	F	F		F	H
	111110	F	H	H		F	H
	111111	H	H	H		H	H

Figure 3. Hermaphrodites (H) and females (F) determined by interaction between nuclear and cytoplasmic genotypes. An individual is female (male sterile) if the cytotype has at least a single 1 matched to a nuclear 0.

ism. This seems sufficiently circular that it should converge on the truth. However, a model based on inferred specificity is almost certain to be the wrong model, so any match of prediction with observation is probably spurious.

The ideal is to build a model of dynamics based on the biochemistry of specificity. That choice is not yet available because the biochemistry is unknown. At this point a reasonable guess about specificity, such as matching alleles, is perhaps more likely to provide a true model than an approach that begins with detectable polymorphism. An approach that insists on starting with detectable polymorphism inevitably leads to introductory statements such as (Connor & Charlesworth 1989): 'The genetics of male-sterility in gynodioecious [mixed hermaphrodite and male-sterile] species has not been easily or often solved ... For studies of gynodioecism the deficiency lies in genetic solutions to actual problems and not in the need for theoretical models'. If only it were so easy! Based on their own genetic solutions they conclude their paper with

> The solution offered here for male-sterility control in *C. selloana* based on up to three complementary loci with recessive alleles, extends the number and kind of control systems for male-sterility in flowering plants. The results suggest, too, that in the original population probably many other loci are segregating for recessive sterility factors. This agreement with the results of Van Damme for dicotyledonous *Plantago lanceolata* is therefore very striking. It is as yet unclear why gynodioecious populations have so many sterility, or restorer, loci polymorphic.

The weakness of these conclusions is typical of many papers that focus exclusively on segregation analysis to determine the genetic control of male sterility. To paraphrase: the genetics are complicated, many loci are involved, and we don't know why.

Theory has provided plausible explanations of polymorphism (Charlesworth 1981; Delannay *et al.* 1981; Gouyon & Couvet 1985; Frank 1989; Gouyon *et al.* 1991). Further progress will probably require biochemical analysis of recognition and specificity between competing nuclear and cytoplasmic genes.

4. BACTERIA AND THEIR VIRAL PARASITES

I have emphasized the importance of recognition and specificity in host–parasite genetics. Bacterial communities are promising model systems because the biochemistry of recognition is relatively easy to study when compared with CMS or plant–pathogen interactions. In this section I describe the natural history of bacterial restriction–modification (R–M) enzymes, which are used to defend against viral attack. I discuss a surprising outcome of bacterial–viral coevolution, in which parasitic attack leads to widespread genetic polymorphism in the bacterial defence system and the near-extinction of the parasites.

Bacteria have a simple recognition-based immunity system that protects them from invasion by foreign DNA (Wilson & Murray 1991). There are two components to the system. Restriction enzymes cut DNA molecules that carry a particular sequence of nucleotides. Modification enzymes recognize the same nucleotide sequence but, instead of cutting the DNA, these enzymes modify the recognition site in a way that protects that molecule from restriction. A bacterial cell's own DNA is modified, otherwise the restriction enzymes would cut the DNA and kill the cell.

R–M enzymes are known for over 200 different recognition sites (Kessler & Manta 1990; Roberts 1990). Circumstantial evidence suggests that defence against bacteriophage viruses has been a powerful force promoting diversity. First, R–M can protect host cells from invading phages (Luria & Human 1952; Arber 1965). Secondly, phages that develop in bacteria with a particular R–M type are modified for the associated recognition sequence. These modified phages can attack other bacteria of the same R–M type, but are sensitive to restriction by different R–M systems. Rare R–M types are favoured because few phages will be modified for their recognition sequence. This frequency-dependent selection promotes diversity of R–M as a defence against phages (Levin 1986, 1988). Thirdly, phages carry a variety of antirestriction mechanisms (Kruger & Bickle 1983; Sharp 1986; Korona *et al.* 1993). For example, many phages lack particular R–M recognition sequences. The probability of having these recognition sequences is very high if no selective pressure is acting on sequence composition.

The circumstantial evidence favours phage-mediated selection as an explanation for R–M diversity. However, direct studies of interactions between phages and bacteria suggest that bacteria resist phage attack by modifying the receptor sites at which phages adsorb and enter the cell (Lenski 1984, 1988; Lenski & Levin 1985). In these studies R–M apparently has little effect on the long-term dynamics of phage and bacteria; this results suggests that R–M diversity may be maintained by processes other than phage-mediated selection in stable communities (Korona & Levin 1993).

Laboratory studies of phages and bacteria maintained in chemostats provide repeatable observations about coevolution between phages and bacteria (Lenski 1988; Korona & Levin 1993). Phages and bacteria are mixed to begin the experiment. No matter what the short-term dynamics, the bacteria usually evolve a set of surface receptors that resist attack by phages. These modified receptors may reduce host growth rate because the receptors used by phages are typically the site for uptake of important nutrients.

With the appearance of receptor-based resistance, the community settles to a balance of resistant bacteria with reduced growth and sensitive bacteria with phage-induced mortality (Levin *et al.* 1977). In these communities, resistant bacteria typically outnumber sensitive bacteria, and there is a small phage population supported on the sensitive strain (Lenski 1988).

The outcome of evolution in laboratory communities can be summarized as follows. If receptor-resistance is rare during the early phases of the experiment, R–M may provide some defence against phage. As the experiment proceeds, receptor-based resistance becomes common, phage become rare, and R–M loses its selective advantage.

The observations from natural populations provide conflicting evidence about the role of R–M. On the one hand, phages are rare in natural isolates (Scarpino 1978) and receptor-based resistance is common (Lenski & Levin 1985). These observations support the view that phage-mediated selection is a very weak force in the maintenance of R–M diversity. On the other hand, phages often carry anti-restriction mechanisms, suggesting that R–M is an important selective force on phage and that, in turn, phages probably influence R–M diversity.

Levin (1986, 1988; Korona & Levin 1993) suggested that the conflicting evidence can be explained by a model in which R–M is advantageous in colonizing new habitats where phages are common and receptor-based resistance for the local phage has not yet evolved. As the newly established community matures, receptor-based resistance spreads and eventually dominates. Thus R–M diversity is maintained by cycles of selection that occur during colonization.

I suggested an alternative model to explain the apparent contradiction that, in some mature communities, phages are rare and R–M systems are diverse (Frank 1994a). My model showed that variation in R–M diversity is itself a direct cause of community structure.

The goal is to explain the community mixture of phages, bacteria with receptor-based resistance to the phages, and bacteria with only R–M defence. Computer studies showed that the abundances of the different phages and bacteria settle to steady values in a mature community, so I describe the results in terms of the community equilibrium.

The main conclusions are shown in figure 4. There are N different bacterial R–M types. Each R–M type has a matching subpopulation of phage that was born in that type; the matching phages have modified DNA that protects them from the defences of that R–M type. The equilibrium abundance of phage modified for each of the R–M types is p^*, thus the total

abundance of phage is Np^*. The abundance of each R–M type is h^*, the total abundance is Nh^*. The abundance of the bacteria with receptor-based resistance to the phage is h_r. The last important parameter is d, which is most strongly affected by the amounts of nutrients available for bacterial growth (Frank 1994a).

The effects of increasing R–M diversity on community composition can be divided into three stages. Each stage is labelled by a circled number in the panels of figure 4.

1. As the number of R–M types increases from $N = 1$, the abundances of phages and phage-sensitive R–M types increase and the abundance of receptor-resistant bacteria decreases. A rare R–M type always invades an equilibrium community with phages because none of the phages are specialized (modified) for the new type. Each new R–M type increases to the point where it maintains its own phage subpopulation that limits the further spread of that R–M type. Phage limitation of each R–M type has two consequences. First, each new R–M type causes an approximately linear increase in the total abundance of phages and R–M types. Second, the resources taken by each new R–M type reduce the abundance of the receptor-resistant population but do not interfere with other R–M types.

2. The receptor-resistant bacteria are eventually driven to extinction when a sufficient number of R–M types have accumulated. Novel R–M types can continue to invade. Each new type causes a reduction in the phage population. This reduction is probably caused by the high proportion of phage deaths that result when the phages invade and are restricted by bacteria for which the phage DNA is unmodified. As N rises, the proportion of bacteria that are resistant to each particular phage increases, $(N - 1)/N$.

3. Further increase in the number of R–M types drives the phages to extinction. At this point, each phage is matched to such a small proportion $(1/N)$ of the bacterial population that phage death rate exceeds the rate of new births. The stable community at the transition between stages 2 and 3 supports a diversity of R–M types but no phages.

The role of habitat quality can be seen by comparing the left panel, with relatively low quality, and the right panel, with relatively high quality.

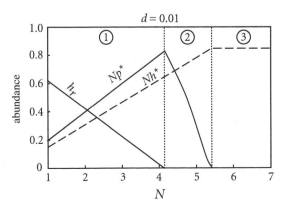
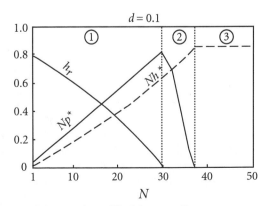

Figure 4. Community composition as a function of the number of R–M types, N.

Richer habitats favour the receptor-resistant bacteria because the competition for nutrients is weaker and the reduced nutrient uptake imposes a lower cost. Thus richer habitats require a greater number of R–M types to drive the receptor-resistant bacteria to extinction.

The interesting aspect of the model is that R–M diversity can have a strong influence on community composition. For a fixed level of nutrients, an increase in R–M diversity causes a shift from a community with few phages and relative dominance by receptor-resistant bacteria to a community in which receptor-based resistance is rare and phages are common. A further increase in R–M diversity can drive the phages to extinction. In a laboratory experiment this sequence would lead to an endpoint that, at first sight, would seem strange: a community of diverse R–M types but an absence of the selective pressure (the phages).

Host diversity drives the parasite to extinction because of the biochemistry of recognition and specificity. Each parasite can attack only $1/N$ of the R–M types. As N increases, the number of hosts available to each phage drops to the point where deaths exceed births, and the phages are driven to extinction.

5. BACTERIOCINS AND ALLELOPATHY

In this section I outline the natural history of another interesting bacterial system. The problem concerns competition between bacterial strains rather than what is usually thought of as a host–parasite interaction. However, the genetic specificity of attack and defence promotes widespread polymorphism in much the same way as in host–parasite systems. Many of the inference problems mentioned above apply to this system, but for the sake of brevity I limit myself to description.

Bacteria often carry plasmids that encode a bacterial toxin (bacteriocin) and immunity to that toxin (Reeves 1972; Hardy 1975; Lewin 1977). Immunity works by neutralizing the toxin after it has entered the cell. Bacteria may also be resistant to bacteriocins because they lack a compatible receptor through which the toxin can enter the cell.

Many distinct bacteriocin types are found within a population. A type is defined by its susceptibility to a set of toxin-producing test strains. With n test strains, there are 2^n possible types. Epidemiological studies frequently use bacteriocin typing to identify and follow pathogenic strains of bacteria. These studies provide information about the diversity of bacteriocin production and susceptibility in populations. For example, Chhibber *et al.* (1988) summarize data on the number of isolates, test strains, and bacteriocin susceptibilities for ten studies of *Klebsiella pneumoniae*. The smallest number of observed types occurred in a study with 200 isolates, four test strains, and 11 types of a possible $2^4 = 16$; the most occurred in a study with 553 isolates, seven test strains, and 64 types of a possible $2^7 = 128$. Similar levels of diversity have been reported for a variety of species (Gaston *et al.* 1989;

Senior & Vörös 1989; Rocha & Uzeda 1990; Traub 1991; Riley & Gordon 1992).

The patterns of resistance and susceptibility are determined by the distribution of toxins, specific immunity, and receptor-based resistance. There appears to be widespread diversity for all three components. Only one model has attempted to explain bacteriocin diversity (Frank 1994*b*). That model emphasized the diversifying effects of varying habitat quality: toxins are relatively more effective in resource-rich habitats. The model provides a basis for further studies, but does not explain the diversity of receptor-based resistance and specific immunity found in natural populations.

Although this system is perhaps the least well understood of my four examples, it may also be the most promising for future studies. Recent molecular work has provided sequences for genes involved in toxin production (Riley 1993*a,b*). As molecular libraries are developed for different toxin, immunity and receptor alleles, it will be possible to measure directly the distribution of polymorphism without the confusion of inferring genotype from complex phenotypic interactions. In addition, laboratory experiments can be used to test predictions about the ecological and genetic interactions that determine the distribution of diversity.

6. DISCUSSION

Coevolutionary systems are difficult to study because the biochemistry of recognition and the associated host–parasite specificity cannot be inferred from phenotypic interactions and detectable polymorphism. For example, in cytoplasmic male sterility the same nuclear polymorphisms may be classified as controlling male sterility without cytoplasmic interaction, as working in pairs to restore fertility for a particular cytoplasm, or as having a one-to-one specificity with a range of cytotypes. These different classifications depend on the cytoplasmic polymorphism available to classify the phenotypes of the nuclear alleles.

Another point of potential confusion is that intense parasite pressure can lead to low frequencies of host resistance. This occurs, for example, in a matching-allele model where each of n host alleles recognizes and resists only $1/n$ of the parasite alleles. An increase in the number of alleles, n, reduces the chance that any individual host will match a particular parasite. Thus parasite pressure causes diversification of both host and parasite, with low frequencies of host resistance and widespread parasitism.

A different outcome occurs when the specificity causes each host type to resist all but one of the parasite types. For example, in the simple bacterial restriction–modification model, each of the N viral types can attack only $1/N$ of the hosts. As the number of types increases, the viral host-range shrinks. If N is sufficiently large, the virus is driven to extinction. Thus parasite pressure causes widespread polymorphism in the host and, eventually, a great reduction in the abundance of parasites.

In each of my four examples I have simplified the natural history and the available theory in order to emphasize these general points about recognition and polymorphism. For example, in restriction–modification systems the viruses have an array of anti-restriction mechanisms, including a lack of the DNA sequence cut by restriction enzymes (Kruger & Bickle 1983; Sharp 1986). In this case the biochemistry of recognition is clear, but very little is known about the distribution of polymorphism in natural populations. Recent surveys by Korona *et al.* (1993) show that natural isolates of the viruses carry a diverse array of defences against restriction. Much more work of this kind is needed, with emphasis on joint studies of biochemistry and natural diversity.

On the theoretical side, there is often a call for greater realism to match the complexity of host–parasite genetics. More realism, by itself, will provide little insight. The simplified theories that I have presented emphasize general trends among these complex systems. The trends may not hold, but they must be replaced by equally simple tendencies if there is to be any general science of host–parasite genetics. Of course, each individual system will have its own peculiar natural history and dynamics that must be studied independently. However, the uniqueness of each system should not obscure the fact that there are only a limited number of ways for hosts to recognize parasites, and only a limited number of patterns for matching host resistance to parasite host-range.

I have ignored evolutionary dynamics and spatial variation in order to focus on recognition and polymorphism. It is clear that non-equilibrium fluctuations in time and space play a crucial role in host–parasite genetics. Many properties of dynamics have been studied with models (Seger 1992). Two predictions may be useful for a wide variety of host–parasite interactions.

First, the capacity of the parasite population for explosive growth is often the most important factor controlling the magnitude of fluctuations in population sizes and in gene frequencies (May & Anderson 1983; Frank 1991*a*, 1993*a*). Fast-growing parasites often cause local epidemics. These epidemics may be started by a limited set of genotypes that happens to avoid the local host resistance. These few pathogen genotypes favour a shift in the host population to resistant genotypes. The net effect is changing population sizes and gene frequencies in each local population, with spatially isolated populations cycling through epidemics and genotypes at different times (Gouyon & Couvet 1985; Frank 1989, 1991*b*, 1993*a*; Burdon *et al.* 1989, 1990; Thompson & Burdon 1992). Locally novel parasite genotypes that cause epidemics are likely to arrive by migration from other patches, and host resistance to counter the new parasites also may be introduced by migration. Thus local extinctions of genotypes and subsequent recolonizations from other patches can drive a continual space-time turnover of gene frequencies. By contrast, slow-growing diseases are likely to have more stable population sizes and gene frequencies, with greater diversity in each population and less spatial variation (Frank 1991*a*, 1993*a*).

The second prediction is that an increasing number of genotypes in a host–parasite interaction causes greater space-time variation (Seger 1988; Frank 1989, 1991*a*; Hamilton *et al.* 1990). For example, in a matching-allele model with only two alternative types ($n = 2$), each local population is likely to have both of the host resistance alleles. This prevents a parasite genotype from spreading rapidly because the host population has a specific resistance factor for all parasite genotypes. By contrast, when there are many allelic variants, a local population of hosts is likely to be missing one or more variants. The matching parasite can invade and spread rapidly, driving out the other parasite genotypes and strongly favouring the introduction and spread of the matching host allele. Thus, the greater the number of possible genotypes, the greater the potential for extensive space-time variation. An increase in the number of possible genotypes may also cause a decline in local variation by enhancing the tendency for rapid turnover of locally successful genotypes.

Each of my examples and general conclusions focuses on host–parasite systems with simple genetic specificities. By contrast, many hosts defend against parasites by inducing a variety of non-specific defences and by acquiring specific recognition with antibody selection. It is difficult to piece together a complete evolutionary story for recognition and polymorphism for these systems because there is rarely sufficient information about both host and parasite. The appeal of my four examples is their simplicity and the natural history data that suggest how host–parasite coevolution works.

There is much interest in the coevolution of different kinds of host–parasite and plant–herbivore interactions (Wakelin & Blackwell 1988; Harvell 1990*a*,*b*; Crawley 1992; Fritz & Simms 1992). However, for these systems it is difficult to achieve even the limited realism of my examples. Some recent models have laid the groundwork for the coevolutionary genetics of quantitative traits (Seger 1992; Saloniemi 1993; Frank 1993*d*, 1994*c*) and systems in which hosts induce non-specific chemical and structural defences in response to attack (Clark & Harvell 1992; Frank 1993*c*). Further progress will benefit greatly from a few systems in which specificity and polymorphism are described for both the host and the parasite.

For comments on the manuscript I thank R. M. Bush and M. A. Riley's discussion group at Yale: M. Feldgarden, Y. Tan, J. Wernegreen and E. Wright. My research is supported by NSF grant DEB-9057331 and NIH grants GM42403 and BRSG-S07-RR07008.

REFERENCES

Arber, W. 1965 Host controlled modification of bacteriophage. *A. Rev. Microbiol.* **19**, 365–368.

Burdon, J.J. 1987 *Diseases and plant population biology.* Cambridge University Press.

Burdon, J.J., Brown, A.H.D. & Jarosz, A.M. 1990 The

spatial scale of genetic interactions in host–pathogen coevolved systems. In *Pests, pathogens and plant communities* (ed. J. J. Burdon & S. R. Leather), pp. 233–247. Oxford: Blackwell Scientific Publications.

Burdon, J.J., Jarosz, A.M. & Kirby, G.C. 1989 Pattern and patchiness in plant–pathogen interactions – causes and consequences. *A. Rev. Ecol. Syst.* **20**, 119–136.

Charlesworth, D. 1981 A further study of the problem of the maintenance of females in gynodioecious species. *Heredity, Lond.* **46**, 27–39.

Chhibber, S., Goel, A., Kapoor, N., Saxena, M. & Vadehra, D.V. 1988 Bacteriocin (klebocin) typing of clinical isolates of *Klebsiella pneumoniae*. *Eur. J. Epidemiol.* **4**, 115–118.

Christ, B.J., Person, C.O. & Pope, D.D. 1987 The genetic determination of variation in pathogenicity. In *Populations of plant pathogens: their dynamics and genetics* (ed. M. S. Wolfe & C. E. Caten), pp. 7–19. Oxford: Blackwell Scientific Publications.

Clark, C.W. & Harvell, C.D. 1992 Inducible defenses and the allocation of resources – a minimal model. *Am. Nat.* **139**, 521–539.

Connor, H.E. & Charlesworth, D. 1989 Genetics of male-sterility in gynodioecious *Cortaderia* (Gramineae). *Heredity, Lond.* **63**, 373–382.

Couvet, D., Atlan, A., Belhassen, E., Gliddon, C., Gouyon, P.H. & Kjellberg, F. 1990 Co-evolution between two symbionts: the case of cytoplasmic male-sterility in higher plants. *Oxf. Surv. Evol. Biol.* **7**, 225–249.

Crawley, M.J. (ed). 1992 *Natural enemies: the population biology of predators, parasites and diseases.* Oxford: Blackwell Scientific.

Delannay, X., Gouyon, P.-H. & Valdeyron, G. 1981 Mathematical study of the evolution of gynodioecy with cytoplasmic inheritance under the effect of a nuclear restorer gene. *Genetics* **99**, 169–181.

Edwardson, J.R. 1970 Cytoplasmic male sterility. *Bot. Rev.* **36**, 341–420.

Fincham, J.R.S., Day, P.R. & Radford, A. 1979 *Fungal genetics*, 4th edn. Berkeley: University of California Press.

Flor, H.H. 1956 The complementary genic systems in flax and flax rust. *Adv. Genet.* **8**, 29–54.

Flor, H.H. 1971 Current status of the gene-for-gene concept. *A. Rev. Phytopathol.* **9**, 275–296.

Frank, S.A. 1989 The evolutionary dynamics of cytoplasmic male sterility. *Am. Nat.* **133**, 345–376.

Frank, S.A. 1991*a* Ecological and genetic models of host–pathogen coevolution. *Heredity, Lond.* **67**, 73–83.

Frank, S.A. 1991*b* Spatial variation in coevolutionary dynamics. *Evol. Ecol.* **5**, 193–217.

Frank, S.A. 1992 Models of plant–pathogen coevolution. *Trends Genet.* **8**, 213–219.

Frank, S.A. 1993*a* Coevolutionary genetics of plants and pathogens. *Evol. Ecol.* **7**, 45–75.

Frank, S.A. 1993*b* Specificity versus detectable polymorphism in host–parasite genetics. *Proc. R. Soc. Lond.* B **254**, 191–197.

Frank, S.A. 1993*c* A model of inducible defense. *Evolution* **47**, 325–327.

Frank, S.A. 1993*d* Evolution of host–parasite diversity. *Evolution.* **47**, 1721–1732.

Frank, S.A. 1994*a* Polymorphism of bacterial restriction-modification systems: the advantage of diversity. *Evolution.* (In the press.)

Frank, S.A. 1994*b* Spatial polymorphism of bacteriocins and other allelopathic traits. *Evol. Ecol.* **8**, 369–386.

Frank, S.A. 1994*c* Coevolutionary genetics of hosts and parasites with quantitative inheritance. *Evol. Ecol.* **8**, 74–94.

Fritz, R.S. & Simms, E.L. 1992 *Plant resistance to herbivores and pathogens.* Chicago: University of Chicago Press.

Gabriel, D.W. & Rolfe, B.G. 1990 Working models of specific recognition in plant-microbe interactions. *A. Rev. Phytopathol.* **28**, 365–391.

Gaston, M.A., Strickland, M.A., Ayling-Smith, B.A. & Pitt, T.L. 1989 Epidemiological typing of *Enterobacter aerogenes*. *J. clin. Microbiol.* **27**, 564–565.

Gouyon, P.-H. & Couvet, D. 1985 Selfish cytoplasm and adaptation: variations in the reproductive system of thyme. In *Structure and functioning of plant populations* (ed. J. Haeck & J. W. Woldendorp), pp. 299–319. New York: North-Holland.

Gouyon, P.-H., Vichot, F. & Van Damme, J.M.M. 1991 Nuclear-cytoplasmic male sterility: single-point equilibria versus limit cycles. *Am. Nat.* **137**, 498–514.

Hamilton, W.D., Axelrod, R. & Tanese, R. 1990 Sexual reproduction as an adaptation to resist parasites (a review). *Proc. natn. Acad. Sci. U.S.A.* **87**, 3566–3573.

Hanson, M.R. 1991 Plant mitochondrial mutations and male sterility. *A. Rev. Genet.* **25**, 461–486.

Hardy, K.G. 1975 Colicinogeny and related phenomena. *Bacteriol. Rev.* **39**, 464–515.

Harvell, C.D. 1990*a* The evolution of inducible defence. *Parasitology* **100**, S53–S61.

Harvell, C.D. 1990*b* The ecology and evolution of inducible defenses. *Q. Rev. Biol.* **65**, 323–340.

Kessler, C. & Manta, V. 1990 Specificity of restriction endonucleases and DNA modification methyltransferases – a review. *Gene* **92**, 1–248.

Koelewijn, H.P. 1993 On the genetics and ecology of sexual reproduction in *Plantago coronopus*. Ph.D. thesis, University of Groningen, The Netherlands.

Korona, R., Korona, B. & Levin, B.R. 1993 Sensitivity of naturally occurring coliphages to Type I and Type II restriction and modification. *J. gen. Microbiol.* **139**, 1283–1290.

Korona, R. & Levin, B.R. 1993 Phage-mediated selection and the evolution and maintenance of restriction-modification. *Evolution* **47**, 556–575.

Kruger, D.H. & Bickle, T.A. 1983 Bacteriophage survival: multiple mechanisms for avoiding deoxyribonucleic acid restriction systems of their hosts. *Microbiol. Rev.* **47**, 345–360.

Laser, K.D. & Lersten, N.R. 1972 Anatomy and cytology of microsporogenesis in cytoplasmic male sterile angiosperms. *Bot. Rev.* **38**, 425–454.

Lawrence, G.J., Mayo, G.M.E. & Shepherd, K.W. 1981 Interactions between genes controlling pathogenicity in the flax rust fungus. *Phytopathology* **71**, 12–19.

Lenski, R.E. 1984 Coevolution of bacteria and phage: are there endless cycles of bacterial defenses and phage counterdefenses? *J. theor. Biol.* **108**, 319–325.

Lenski, R.E. 1988 Dynamics of interactions between bacteria and virulent bacteriophage. *Adv. microb. Ecol.* **10**, 1–44.

Lenski, R.E. & Levin, B.R. 1985 Constraints on the coevolution of bacteria and virulent phage: a model, some experiments, and predictions for natural communities. *Am. Nat.* **125**, 585–602.

Leonard, K.J. & Czochor, R.J. 1980 Theory of genetic interactions among populations of plants and their pathogens. *A. Rev. Phytopathol.* **18**, 237–258.

Levin, B.R. 1986 Restriction-modification immunity and the maintenance of genetic diversity in bacterial populations. In *Evolutionary processes and evolutionary theory* (ed. S. Karlin and E. Nevo), pp. 669–688. New York: Academic Press.

Levin, B.R. 1988 Frequency-dependent selection in

bacterial populations. *Phil. Trans. R. Soc. Lond.* B **319**, 459–472.

Levin, B.R., Stewart, F.M. & Chao, L. 1977 Resource-limited growth, competition, and predation: a model and experimental studies with bacteria and bacteriophage. *Am. Nat.* **111**, 3–24.

Levin, S.A. 1983 Some approaches to the modelling of coevolutionary interactions. In *Coevolution* (ed. M. H. Nitecki), pp. 21–65. Chicago: University of Chicago Press.

Lewin, B. 1977 *Gene expression, (Plasmids and phages).* New York: Wiley.

Lewis, D. 1941 Male sterility in natural populations of hermaphrodite plants: the equilibrium between females and hermaphrodites to be expected with different types of inheritance. *New Phytol.* **40**, 56–63.

Luria, S.E. & Human, M.L. 1952 A non-hereditary host-induced variation in bacterial viruses. *J. Bact.* **64**, 557–559.

May, R.M. & Anderson, R.M. 1983 Epidemiology and genetics in the coevolution of parasites and hosts. *Proc. R. Soc. Lond.* B **219**, 281–313.

Person, C. & Mayo, G.M.E. 1974 Genetic limitations on models of specific interactions between a host and its parasite. *Can. J. Bot.* **52**, 1339–1347.

Potts, W.K. & Wakeland, E.K. 1993 Evolution of MHC genetic diversity: a tale of incest, pestilence and sexual preference. *Trends Genet.* **9**, 408–412.

Reeves, P. 1972 *The bacteriocins.* New York: Springer-Verlag.

Richards, A.J. 1986 *Plant breeding systems.* London: Unwin Hyman.

Riley, M.A. 1993*a* Positive selection for colicin diversity in bacteria. *Molec. Biol. Evol.* **10**, 1048–1059.

Riley, M.A. 1993*b* Molecular mechanism of colicin evolution. *Molec. Biol. Evol.* **10**, 1380–1395.

Riley, M.A. & Gordon, D.M. 1992 A survey of Col plasmids in natural isolates of *Escherichia coli* and an investigation into the stability of Col-plasmid lineages. *J. gen. Microbiol.* **138**, 1345–1352.

Roberts, R.J. 1990 Restriction enzymes and their isoschizomers. *Nucl. Acids Res.* **18** (Suppl.), 2331–2365.

Rocha, E.R. & de Uzeda, M. 1990 Antagonism among *Bacteriodes fragilis* group strains isolated from middle ear exudates from patients with chronic suppurative otitis media. *Ear Nose Throat J.* **69**, 614–618.

Saloniemi, I. 1993 A coevolutionary predator-prey model with quantitative characters. *Am. Nat.* **141**, 880–896.

Scarpino, P.V. 1978 Bacteriophage indicators. In *Indicators of viruses in water and food* (ed. G. Berg), pp. 201–227. Ann Arbor, Michigan: Ann Arbor Science.

Seger, J. 1988 Dynamics of some simple host–parasite models with more than two genotypes in each species. *Phil. Trans. R. Soc. Lond.* B **319**, 541–555.

Seger, J. 1992 Evolution of exploiter-victim relationships. In *Natural enemies: the population biology of predators, parasites and diseases* (ed. M. J. Crawley), pp. 3–25. Oxford: Blackwell Scientific.

Senior, B.W. & Vörös, S. 1989 Discovery of new morganocin types of *Morganella morganii* in strains of diverse serotype and the apparent independence of bacteriocin type from serotype of strains. *J. med. Microbiol.* **29**, 89–93.

Sharp, P.M. 1986 Molecular evolution of bacteriophages: evidence of selection against the recognition sites of host restriction enzymes. *Molec. Biol. Evol.* **3**, 75–83.

Thompson, J.N. & Burdon, J.J. 1992 Gene-for-gene coevolution between plants and parasites. *Nature, Lond.* **360**, 121–125.

Traub, W.H. 1991 Bacteriocin typing and biotyping of clinical isolates of *Serratia marcescens. Int. J. med. Microbiol.* **275**, 474–486.

Vanderplank, J.E. 1968 *Disease resistance in plants.* New York: Academic Press.

Wakelin, D.M. & Blackwell, J.M. 1988 *Genetics of resistance to bacterial and parasitic infection.* London: Taylor & Francis.

Wilson, G.G. & Murray, N.E. 1991 Restriction and modification systems. *A. Rev. Genet.* **25**, 585–627.

3

Viral pathogens and the advantage of sex in the perennial grass *Anthoxanthum odoratum*

STEVEN E. KELLEY

Department of Biology, Emory University, 1510 Clifton Road, Atlanta, Georgia 30322, U.S.A.

SUMMARY

The ubiquity of sexual reproduction among plants and animals remains one of the major unresolved paradoxes of modern evolutionary biology. In order for sex to be maintained in populations, sex must confer immediate and substantial fitness benefits. Theoreticians have proposed numerous mechanisms to explain how such advantages arise, but experimental data are few. In one well-studied population of the perennial grass *Anthoxanthum odoratum* in a mown North Carolina field, sexual offspring have been found to have significantly higher fitness than asexual offspring. More recent field experiments show that an aphid-transmitted virus, barley yellow dwarf (BYDV)-strain SGV, specifically transmitted by *Schizaphus graminum*, frequently infects *Anthoxanthum* progeny soon after transplantation into the field. BYDV infection is asymptomatic in *Anthoxanthum*, but BYDV-inoculated clones planted directly in the field had significantly lower fitness than healthy controls.

Sexual females have been hypothesized to gain a fitness advantage for their offspring in the presence of pathogens either by providing 'an escape in time' from pathogens preadapted to the parental genotype or through the production of rare genotypes, which escape frequency-dependent infection. When parental clones and seed-derived sexual offspring were planted in identical but separate arrays in sites near where the parent was collected, parental clones were twice as frequently infected as sexual offspring. Factors other than genetic variation may have contributed to differences in levels of infection between sexual and asexual progeny: in this experiment, clonally derived asexual offspring tillers were slightly larger than seed-derived sexual tillers; in field experiments, larger plants were more frequently infected than smaller plants. When different families were planted into a common site, there was evidence that genotypes were less frequently infected when locally rare than when common.

Taken together, the data suggest that BYDV infection generates advantages for rare or sexually produced genotypes in *Anthoxanthum*. The pattern of infection is likely to result from a complex interaction between vector, host, and viral genetics and population structure, vector behaviour, and host and vector dispersal patterns. Sexually produced genotypes appear to benefit because they are both novel and rare, but the observed minority advantage was weak. Other viral, bacterial, and fungal pathogens in this *Anthoxanthum* population may act as frequency-dependent selective forces in different places in the field, collectively generating the substantial and observed overall fitness advantage of rare genotypes. Further study is needed to elucidate their role. Nevertheless, the data do show that viral pathogens, which are often asymptomatic, play a significant evolutionary role in plant populations.

1. INTRODUCTION

Viruses are ubiquitous in wild plant populations. Surveys of wild populations suggest that an average of 10% of individuals in a population are infected (Kelley 1993). But epidemics (in which up to 67% of individuals are infected) are regularly found in small surveys of multiple populations (Kelley 1993). When plants are in their natural environments, it is commonplace for virus replication to occur in individuals that show no detectable abnormalities (Cooper & MacCallum 1984). Viruses are difficult to detect; because the absence of symptoms is often assumed to indicate little fitness cost to infected hosts,

it is not surprising that field ecologists often overlook viruses in plant populations.

Since the discovery of an 'intrinsic cost' of sex (Williams 1975; Maynard Smith 1978; Bell 1982), there has been considerable interest in the adaptive significance of sexual reproduction. Theoreticians have proposed that pathogens represent the major selective force favouring the maintenance of genetic variation and sexual reproduction in plant as well as animal populations (Jaenike 1978; Hamilton 1980, 1990; Hutson & Law 1981; Tooby 1982). According to theory, to be maintained in populations, sexual reproduction must confer immediate, substantial (twofold) fitness benefits. The discovery of the cost of sex has spawned a number of experimental studies (Antonovics & Ellstrand 1984; Ellstrand & Antonovics 1985; Bierzychudek 1987; Lively 1987; Kelley *et al.* 1988; Kelley 1989*a,b*; Lively *et al.* 1990).

One of the most well-studied systems used to test hypotheses for the adaptive significance of sexual reproduction is a population of the short-lived perennial bunch grass *Anthoxanthum odoratum* growing in a mown North Carolina field (Antonovics & Ellstrand 1984; Ellstrand & Antonovics 1985; Kelley *et al.* 1988; Kelley 1989*a,b*, 1993). In these studies, seeds taken from single inflorescences were germinated, grown to large plants in a glasshouse, and then divided into single tillers to constitute 'sexual progeny'. Vegetative ramets sampled from adults were taken to the glasshouse, grown to large size, and likewise divided into single tillers to constitute 'asexual progeny'. Fitness was assessed as the net reproductive rate, estimated as counts of the number of inflorescences summed over the lifetime of the progeny. The number of inflorescences has been shown to be highly correlated with the number of spikelets, and hence male and female contribution to fitness, there being one seed and two anthers per spikelet. When sexual and asexual progeny were planted directly into the field in arrays that simulated the natural dispersal profile around adults, sexual progeny were more fit than asexual progeny by a factor of 1.55 (Kelley *et al.* 1988).

The pattern of the advantage of sex in numerous experiments suggested that predators or pathogens were responsible for this advantage, but there was little obvious evidence of pathogen or predator attack. However, aphids (*Schizaphus graminum*) were occasionally observed, and in one experiment, the presence of aphids was correlated with advantages for genetic variability in seedlings (Schmitt & Antonovics 1986).

In an initial survey, I tested 45 flowering *Anthoxanthum* plants in this field for a cosmopolitan, aphid-borne RNA virus, barley yellow dwarf luteovirus (Rochow & Duffus 1981), by using direct enzyme-linked immunosorbent assay (ELISA). Twenty percent of adult *Anthoxanthum* tested positive for BYDV-strain SGV (transmitted specifically by *S. graminum* (Kennedy *et al.* 1962)) and 2.2% for BYDV-strain RPV (transmitted specifically by *Rhopolosiphum padi* (Kennedy *et al.* 1962)). BYDV induces yellowing, reddening and brittleness of leaves, and stunting in many crop hosts, but is asymptomatic in *Anthoxanthum* (Kurstak 1981). The virus is phloem-restricted and cannot be spread through contact or through seed. Rather, BYDV consists of multiple different strains, which are specifically transmitted by different aphid species. The virus is transmitted in a persistent manner, and extended feeding on infected plants is necessary in order for aphids to acquire the ability to transmit the virus.

2. PATHOGEN SPREAD IN THE FIELD: THE 'ESCAPE IN TIME' HYPOTHESIS

Pathogen infection can explain short-term advantages of sex if novel genotypes produced by sexual females 'escape in time' from pathogens that are preadapted

to or spread preferentially among clones of the parent (Bremermann 1980; Rice 1983). The hypothesis predicts that if sexual and asexual offspring are planted near the parent, asexual offspring will be more frequently infected than sexual offspring.

To test this prediction, both sexual and asexual progeny were generated for each of 18 adult plants sampled for single tillers and several inflorescences in May 1991. Nine parents were sampled at regular 4 m intervals in each of two transects (northern, N, and southern, S) across the central part of the field (figure 1). Tillers sampled from parents were grown in the glasshouse and divided into single tillers to generate 'asexual progeny'. Seeds from the same parents germinated in flats in August 1991 were grown in the glasshouse and divided into single tillers to generate 'sexual progeny'. Because *Anthoxanthum* is self-incompatible and wind-pollinated, individuals deriving from seeds were likely to be highly genetically variable, sometimes full sibs of each other but mostly half-sibs (Kelley *et al.* 1988).

Both sexual and asexual progeny were planted in the field, with minimal disturbance of the vegetation, in 'home sites' (within 1.5 m of the location at which the parent was sampled). All progeny were planted into hexagonal arrays at a fixed density (4.0 cm between adjacent plants) at which advantages for sex were large in previous experiments (Ellstrand & Antonovics 1985; Kelley *et al.* 1988).

Asexual offspring were planted in a single hexagonal design with six clones surrounding a central clone.

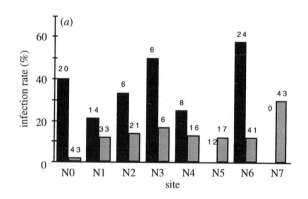

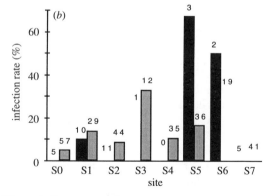

Figure 1. Percentage of asexual (solid) and sexual (stippled) offspring planted into home sites infected with BYDV-strain SGV. Number of plants alive and tested at each site are shown above the bars. Letters beneath the bars designate the different home sites. See text for further details.

Sexual progeny were planted into three designs, which differed in the amount of local genetic variability. Sexual offspring were planted with a single clone of each (seed-derived) genotype planted at the centre surrounded in separate hexagonal plantings with: (*a*) six clones of the same genotype, or (*b*) six maternal half-sib sexual progeny, or (*c*) six unrelated sexual progeny. Each of the three sexual treatments was replicated once each for eight unique sexual genotypes obtained from the female parent sampled at that site. Thus, the three sexual treatments (*a–c*) were each replicated eight times for each parent (if tillers were available) and the asexual treatments were replicated six times. A total of 3255 ramets was planted on 22–30 November 1991.

All material was serologically tested for BYDV-strain SGV, immediately before planting the experiment, by using direct ELISA with sera supplied by AGDIA (Elkhart, Indiana, U.S.A.). Four of 18 parental clones had been naturally infected in the field before sampling and tested positive for BYDV-strain SGV (N2, N3, S4, S6). Tillers were weighed before planting.

A frost immediately after transplantation killed many plants. With few individuals alive in each treatment, there was little difference among the three sexual treatments, since plants had few neighbours. The three sexual treatments were therefore considered as a single treatment in statistical analyses. All plants were surveyed on 3–7 June 1992, the number of vegetative and reproductive tillers were counted, and a small sample of leaf was taken from each living plant for serological testing.

Sexual and asexual progeny planted into home sites were tested for BYDV-strain SGV. The results showed that sexual and asexual progeny were rapidly and frequently infected after transplantation into home field sites (figure 1). Asexual progeny were more than twice as frequently infected as sexual progeny. Overall, 28.3% of asexual progeny and 11.4% of sexual progeny were infected with BYDV-strain SGV (difference: $G = 20.2$, d.f. $= 1$, $p < 0.001$). Excluding parental clones that were infected before the experiment, 26.5% of asexual and 11.7% of sexual progeny tested positive for BYDV-strain SGV (difference: $G = 13.2$, d.f. $= 1$, $p < 0.001$). Local 'hot spots' of infection occurred within short distances, with infection rates as high as 67%. Excluding the sites in which parental clones were infected before the experiment, there was significant among-site variation in the frequency of infection ($G = 71.3$, d.f. $= 22$, $p < 0.001$) and among-site variation in the relative frequency of infection of asexual and sexual progeny ($G = 25.2$, d.f. $= 11$, $p < 0.01$). Although these among-site differences in the infection rates were significant, in almost all cases, asexual progeny were more frequently infected than sexual progeny.

At the start of the experiment, there were slight size differences between asexual and sexual tillers, although these were not statistically significant (asexual tiller mean mass 0.60 g, sexual tiller mean mass 0.47 g, ANOVA square-root transformed data $f = 3.34$, d.f. $= 1, 208$, $p < 0.07$).

3. PATHOGEN SPREAD IN THE FIELD: THE 'MINORITY ADVANTAGE HYPOTHESIS'

According to the 'minority advantage' or Red Queen hypothesis, sex will be favoured if rare genotypes produced by sexual females escape pathogens that are spread in a frequency-dependent manner. It was not possible to test this hypothesis in the earlier experiment, because the frost eliminated differences between 'minority' and 'majority' treatments. Consequently, a second experiment was planted near the first, using the seeds of nine parents (described above) germinated in the glasshouse, grown, and then divided into single tillers. Because BYDV is not seed-transmitted, this procedure ensured that all starting material was free of virus infection. In each of five sites, located at 2 m intervals, genotypes from each of the nine families were represented in three treatments with a genotype planted in hexagonal arrays (4.0 cm spacing) surrounded by (*a*) identical clones, (*b*) maternal half-sibs, and (*c*) unrelated individuals. The minority advantage hypothesis predicts that virus infection will be more frequent in treatments with less genetic variability. All planting material was tested for BYDV-strain SGV before planting the experiment (to ensure that plants were virus-free). A total of 945 tillers were planted on 21–23 December 1991 and surveyed during the period 20 May–2 June 1992, at which time a small leaf sample was taken and serologically tested for BYDV-strain SGV by using direct ELISA with sera supplied by AGDIA (Elkhart, Indiana, U.S.A.).

After approximately six months in the field in one site (site 1), 29% of the plants became infected with BYDV-strain SGV (figure 2) and genotypes were 1.63 times less likely to be infected when rare (unrelated treatment) than when common (identical treatment). At the other four sites, the incidence was small, with

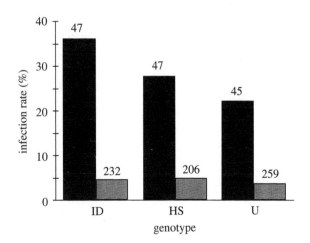

Figure 2. Frequency of BYDV-strain SGV infection in the three local variability treatments shown for site 1 (solid bars) and sites 2–5 (stippled bars) separately. ID, a genotype planted in the centre of a hexagon surrounded by six clonal copies; HS, the same genotype planted in the centre surrounded by six maternal half-sibs; U, the same genotype planted in the centre surrounded with six unrelated individuals. Numbers of plants alive and serologically tested are shown above the bars.

only slightly more plants infected in the less compared with the more genetically variable plantings; no minority advantage was seen. Although minority advantage was seen in one site, these differences in infection frequency were not statistically significant, perhaps because of the small sample size (site × treatment × infection frequency interaction $G = 4.8$, d.f. $= 8$, n.s.; site × infection frequency interaction $G = 56.8$, d.f. $= 12$, $p < 0.01$; treatment × infection frequency interaction $G = 6.0$, d.f. $= 10$, n.s.).

This experiment was repeated in the following year at an adjacent site in the field. Preliminary results showed that minority advantage was observed in four of six sites in the field, with statistically significant heterogeneity found in one of those sites (data unpublished).

4. FITNESS CONSEQUENCES OF BYDV INFECTION

For BYDV infection to act as an important selective force favouring sex, viral infection must reduce individual fitness. The fitness consequences of BYDV-strain SGV infection can be estimated by comparing the performance of cloned parental (asexual progeny) and (sexual) progeny tillers used in the 'escape in time' experiment separately for (i) parents that had been naturally infected in the field with BYDV-strain SGV (before clonal propagation), and (b) parents that were not infected (tables 1 and 2).

When planted into sites close to where parents were sampled (home sites), asexual offspring that were initially free of BYDV infection had 26% higher fecundity than sexual progeny of those same parents. However, asexual progeny that were initially infected (because they had been vegetatively propagated from BYDV-infected parents) had almost fourfold lower fecundity than seed-derived sexual progeny, when

Table 1. *The fitness consequences of virus infection*

(The relative fitnesses of asexual and sexual offspring are shown separately for sites in which all asexual offspring were BYDV-infected because they were clonally propagated from virus-infected parents, and for sites in which asexual offspring were initially free of BYDV-infection. Differences were tested for statistical significance by using a *t*-test, which assumed different variances.)

Parents uninfected with BYDV-SGV	Asexual offspring ($n = 118$)	Sexual offspring ($n = 429$)
vegetative size	2.25	2.16
fecundity	1.08	0.86**
Parents initially infected with BYDV	Asexual offspring ($n = 14$)	Sexual offspring ($n = 90$)
vegetative size	1.71	1.77
fecundity	0.14	0.51*

Symbols: **, $p < 0.025$; *, $p < 0.05$.

Table 2. *Patterns of infection*

(Data are vegetative and reproductive tiller number for plants infected during the first six months of the experiment. Differences between the means were tested for statistical significance by using a *t*-test, which assumed different variances.)

	BYDV serological test result	
	positive ($n = 38$)	negative ($n = 95$)
vegetative size	2.37	1.99^{+}
fecundity	1.47	1.15*

Symbols: $^{+}$, $p < 0.06$; *, $p < 0.05$.

both were planted into sites close to where their parents were sampled. These differences were highly statistically significant.

Because these parental clones had been infected in the field, it was possible that these were less fit genotypes (using 'fitness' in its everyday sense), which had been more susceptible to infection. However, plants that became infected during these field experiments were consistently larger and manifestly more fit than those that did not become infected (table 2). Genotypes with initially higher fitness appeared to be more prone to infection than less fit genotypes. These data seem to indicate that, in the absence of pathogen attack, asexual offspring were slightly more fit than sexual offspring in local home sites.

To measure the fitness consequences of BYDV infection directly, nine seeds obtained from the females sampled in this field were germinated in August 1993, grown to large size, and then divided into single tillers. Three tillers of each genotype were individually inoculated with BYDV-strain SGV by placing seven aphids (*Schizaphus graminum* obtained from field collections and cultured in the glasshouse on a BYDV-SGV infected *Anthoxanthum* plant) on each tiller and allowing them to feed for ten days. The BYDV-infected *Anthoxanthum* used as a source for virus infection was originally sampled from the North Carolina field, and then vegetatively propagated in the glasshouse. To serve as negative controls, single tillers of *Anthoxanthum* were mock-inoculated by placing seven virus-free individuals of *S. graminum* (cultured in the glasshouse on virus-free *Anthoxanthum* plants) onto plants and allowing them to feed for ten days. To serve as an additional negative control, single tillers were left uninoculated. After feeding, aphids were removed from individual plants in the inoculation and mock-inoculation treatments, and two weeks later all plants were tested for BYDV-SGV infection by using direct ELISA. Plants that tested positive, along with negative controls, were propagated for three months in the glasshouse, then divided into single tillers, individually weighed, and planted directly into the field in a randomized block design on 15–18 March 1994. All individuals are surveyed on 15 May 1994, at which time the number of vegetative and reproductive tillers were counted to provide estimates of vegetative size and fecundity, respectively.

Preliminary results from this experiment showed that, although there was no significant size difference between BYDV-infected and healthy tillers at the time of planting, BYDV-infected plants were 12% smaller (mean size, infected:healthy plants 1.7:1.9 tillers, $t = 2.12$, d.f. 341, $p < 0.025$) and had 80% fewer inflorescences than healthy plants (infected:healthy 0.006:0.03 mean inflorescence number, $t = 1.62$, d.f. $= 254$, $p < 0.06$). Thus although growth of the host may be reduced only a little by the pathogen, the effect of the latter on its host's genetic fitness is quite severe.

5. DISCUSSION

Other studies of the consequences of sex, in a dioecious freshwater snail with sexual and parthenogenetic populations (Lively 1987) and coexisting sexual and clonal fish (Lively *et al.* 1990) have implicated parasites in favouring genetic diversity. More recently, studies of a perennial herb, *Eupatorium chinense* (Yahara & Oyama 1993) found considerably higher rates of infection by the geminivirus tobacco leaf curl in agamospermous polyploid populations (30% plants) than in related sexual diploid populations (3%). The results here directly implicate a pathogenic RNA virus as a factor favouring sex in a plant population. The idea that parasites (pathogens) are important to the maintenance of sexual reproduction appears to be quite general.

Despite the intensity with which this *Anthoxanthum* population has been studied and the apparent ubiquity of BYDV infection, BYDV managed to escape detection because it caused no overt foliar symptoms. The absence of symptoms was not equivalent to the absence of fitness consequences. BYDV significantly reduced both fecundity and growth. Studies of other natural virus–host combinations have shown an association between viral infection and reduced growth and fecundity. MacKenzie (1985) found that *Primula vulgaris* clones infected with arabis mosaic virus had significantly lower growth and survivorship than uninfected clones when both were transplanted back into the field. Yahara & Oyama (1993) found that *Eupatorium chinense* plants naturally infected with tobacco leaf curl geminivirus had significantly lower growth rates and higher mortalities and produced significantly few seeds than uninfected plants. In the first study to directly compare the fitness of healthy and virally infected plants, Kelley (1993) found that clones of *Anthoxanthum odoratum* inoculated with brome mosaic bromovirus had significantly lower growth and fecundity, and higher mortality rates, than healthy controls. Plants inoculated with *Anthoxanthum* latent blanching hordeivirus had lower growth and higher mortality, but had slightly higher fecundity than uninfected controls. The fitness consequences of virus infection will depend on the specific host–viral strain combination.

The results of earlier studies that showed an association between viral infection and lower fitness might have been caused by the increased susceptibility of genotypes of low fitness to infection. In particular,

one of the major factors proposed to explain advantages to sexual reproduction is mutational damage, whereby sex functions as a DNA repair mechanism (Michod & Levin 1988) or as the provider of an efficient elimination of deleterious mutations that have to be judged by their effects (not being detectable for repair) (Kondrashov 1984, 1988). According to this hypothesis, plants with mutational damage (clonally propagated asexual genotypes) will be less robust and will be more susceptible to viral infection. However, in this study, *Anthoxanthum* plants that became infected were significantly more robust (larger) than plants not infected close to the time of infection. Further, minority advantage (escape from pathogens) was found here in experiments in which all individuals were (sexually) seed-derived. Interestingly, Oyama and Yahara (1993) also found that most fit individuals were preferred by the whitefly vector of tobacco leaf curl geminivirus and that these plants were preferentially infected. Viral infection may thus primarily affect the fitness of Sisyphean genotypes (Williams 1975).

Barley yellow dwarf virus is restricted to plant phloem, where it is thought to clog phloem and restrict the flow of metabolites to plant roots (Goodman *et al.* 1986). The physiological basis for the fitness reduction in *Anthoxanthum* is not yet understood and is currently under investigation in our laboratory. However, our studies have revealed that aphids (*S. graminum*) can detect the presence of virus in infected plants and, when given a choice, prefer to feed on BYDV-SGV-infected compared with mock-inoculated plants (Kirkley 1993). Aphid colonies initiated on BYDV-infected, mock-inoculated, or healthy tillers of *Anthoxanthum* grew the most rapidly on infected plants (S. E. Kelley, unpublished data). The fitness reduction observed in the field may thus result both from direct physiological debilitation and from that caused by aphid feeding.

Interactive effects of viral infection and aphid behaviour on plant host fitness may be quite general. Blua *et al.* (1994), using an unrelated aphid–host–virus system (*Aphis gossypii* feeding on *Cucurbita pepo* infected with zucchini yellow mosaic potyvirus) also found enhancement of aphid colony development, which they attributed to increased amino acid concentrations in the phloem of virus-infected plants.

When novel sexual genotypes and old asexual genotypes were planted in sites close to where the parent had been sampled, asexual genotypes were more frequently infected. This home-site disadvantage for asexual offspring may have been caused by the fact that clonally propagated 'asexual' tillers were slightly larger than seed-derived 'sexual' tillers. In our laboratory studies, when given a choice between large and small plants, aphids preferred to feed on larger plants; the degree of preference was related to the size discrepancy between plants (Kirkley 1993). Further, in field experiments, larger plants were more frequently infected than smaller plants. Alternatively, feeding preference of local aphid clones for cloned parental genotypes may have resulted in asexual progeny being more frequently infected than sexual progeny. However, our laboratory studies of aphid movements showed no preferential movement toward particular host-plant genotypes (Kirkley 1993). These studies did not exclude the possibility that aphid feeding preferences might be expressed during the period of probing or through differential feeding durations, which would be of critical importance in determining whether plants became infected. Feeding preference of different aphid clones needs further study. A difficulty of the hypothesis that aphid preference is responsible for the 'escape in time' is that the preference for cloned parental over sexual tillers was seen for parents that were initially uninfected. These parents may have initially escaped infection in their home sites because they were resistant to feeding by the local aphid clones or resistant to infection by the local BYDV variants. In such a case, novel sexually produced genotypes might be more vulnerable to aphid feeding and viral infection.

In some sites, genotypes were less frequently infected when locally rare than when locally common. Frequency-dependent infection may also partly explain the disadvantage for asexual genotypes in home sites. Power (1991) noted that rates of aphid movement (*Rhopolosiphum padi* and *Sitobion avenae*) increased in more genotypically diverse plantings of barley varieties compared with less diverse ones, and that symptoms of BYDV infection rates were consequently reduced. A similar mechanism may be operating for *S. graminum* on *Anthoxanthum*, but it needs to be explicitly demonstrated in this system (and its adaptive significance explored).

In experiments reported here, minority advantage was evident only in a few sites, yet the observed advantage of sex in this population is widely observed (Kelley *et al.* 1988). Other viral, bacterial, and fungal pathogens are certainly acting as selective forces and, if the findings for BYDV are typical, are likely to be favouring rare or novel genotypes. J. Bever (personal communication) has recently found that live soils (from this North Carolina field) cultured by different plant species are often pathogenic for those same plant species in the subsequent generations. Kristi Westover (in my laboratory) has found that plant species pairs generally perform less well when planted in soils from beneath the same than different species combinations, and that this effect is at least in part attributable to the living soil components. Other viruses may be spread in *Anthoxanthum* which heretofore have escaped detection. Experiments are planned to assay for other possible sources of enhanced fitness advantage for sexual compared with asexual progeny.

In addition to vector behavior, host passaging effects may explain the observed patterns of infection in both experiments. Barley yellow dwarf luteoviruses (and the majority of plant viruses) have single-stranded RNA genomes. Single-stranded RNA genomes are known to have extremely high mutation rates (on the order of 1000-fold higher than eukaryotic genomes) because of the absence of a proofreading mechanism (Steinhauer & Holland

1987). Selection on such variation occurring during replication over a short or extended period within one host, or through passaging among hosts, has been repeatedly shown to lead to changes in the host range, vector properties, symptom expression (generally the measure of virulence), within-host virion concentration, or infectivity of virus strains (Dawson 1967; Chiko 1984; Shepherd *et al.* 1987; Hajimorad *et al.* 1991; reviewed in Yarwood 1979). In the course of infecting individual hosts, or through passaging through particular host genotypes, BYDV may evolve to infect particular genotypes more frequently, or cause more severe fitness reductions in host genotypes. The consequences of host passaging is currently under investigation in our laboratory.

Because BYDV is not seed-transmitted, it is clear that reproduction via seed will generate immediate and substantial fitness advantages for seed-derived offspring of BYDV-infected parents, but presumably these advantages would accrue to both sexual and apomictic seed offspring. However, because in the *Anthoxanthum* experimental system asexual offspring are derived through clonal propagation, these offspring would carry the virus if the parent had been naturally infected in the field. In past experiments, this may have contributed to the magnitude of the advantage for sexual offspring that was unrelated to the issue of sexual reproduction and the maintenance of genetic variability. Nevertheless, advantages of genetic variability occur in *Anthoxanthum*. Several experiments have shown advantages for rarity in *Anthoxanthum* using only seed-derived planting material (Antonovics & Ellstrand 1984; Schmitt & Antonovics 1986; see above). Further, advantages for sexual offspring in experiments (Kelley *et al.* 1988) occurred in 21 of 25 sites. Infection rates of flowering adults in this field are typically 20% or less and so carry-over of virus infection is not likely to fully explain the advantage of seed-derived sexual offspring in this system. Nevertheless, an improved estimate of the relative fitness of sex could be obtained in an experiment with a two-generation design: the first generation used to produce virus-free parents, and the second generation of seed-derived sexual and clonally derived asexual offspring planted directly into the field in designs stimulating a natural dispersal profile.

Ecologists have tended to overlook the consequences of RNA viruses for plant populations. Yet, in this study, numerous epidemics of virus infection occurred quickly, over short distances, with significant fitness consequences for the host population, and with no obvious signs of pathogen or predator attack. Viruses appear to be pervasive in plant populations (MacClement & Richards 1956; Barnett & Gibson 1975; Hammond 1981), have significant fitness consequences (Gibbs 1980; MacKenzie 1985), and have the potential for rapid mutation and evolution which may enable them to readily adapt to new hosts (Holland *et al.* 1982; Zimmern 1988). Clearly, RNA viruses in plants represent a good model system to investigate the evolutionary and ecological role of pathogens in natural communities.

I acknowledge the encouragement of J. Antonovics, J. L. Harper, W. D. Hamilton and J. Thompson; the training provided by J. I. Cooper, B. and D. Charlesworth; the helpful comments of B. Levin, S. Schrag, and D. Sims; and the financial support provided by the Royal Society and NSF Grant DEB 9221077.

REFERENCES

Antonovics, J. & Ellstrand, N.C. 1984 Experimental studies of the evolutionary significance of sexual reproduction. I. A test of the frequency-dependent selection hypothesis. *Evolution* **38**, 103–115.

Barnett, O.W. & Gibson, P.B. 1975 Identification and prevalence of white clover viruses and the resistance of *Trifolium* species to these viruses. *Crop Sci.* **15**, 32–37.

Bell, G. 1982 *The masterpiece of nature: the evolution and genetics of sexuality*. Berkeley: University of California Press.

Bierzychudek, P. 1987 Resolving the paradox of sexual reproduction: a review of experimental tests. In *Evolution of sex and its consequences* (ed. S. C. Stearns), pp. 163–174. Basel: Birkhauser.

Blua, M.J., Perring, T.M. & Madore, M.A. 1994 Plant virus-induced changes in aphid population development and temporal fluctuations in plant nutrients. *J. Chem. Ecol.* **20**, 691–707.

Bremermann, H.J. 1980 Sex and polymorphism as strategies in host-pathogen interactions. *J. theor. Biol.* **87**, 671–702.

Chiko, A. 1984 Increased virulence of barley stripe mosaic virus for wild oats: evidence of strain selection by host passage. *Phytopathology* **74**, 595–599.

Cooper, J.I. & MacCallum, F.O. 1984 *Viruses and the environment*. London: Chapman & Hall.

Dawson, J.R.O. 1967 The adaptation of tomato mosaic virus to resistant tomato plants. *Ann. appl. Biol.* **60**, 209–214.

Ellstrand, J. & Antonovics, J. 1985 Experimental studies of the evolutionary significance of sexual reproduction. II. A test of the density dependent selection hypothesis. *Evolution* **39**, 657–666.

Gibbs, A.J. 1980 A plant virus that partially protects its wild legume host against herbivores. *Intervirology* **13**, 42–47.

Goodman, R.N., Kiraly, Z. & Wood, K.R. 1986 *The biochemistry and physiology of plant disease*. Columbia: University of Missouri Press.

Hajimorad, M.R., Kurath, G., Randles, J.W. & Francki, R.I.B. 1991 Change in phenotype and encapsidated RNA segments of an isolate of alfalfa mosaic virus: an influence of host passage. *J. gen. Virol.* **72**, 2885–2893.

Hamilton, W.D. 1980 Sex versus non-sex versus parasite. *Oikos* **35**, 282–290.

Hamilton, W.D. 1990 Sexual reproduction as an adaptation to resist parasites (a review). *Proc. natn. Acad. Sci. U.S.A.* **87**, 3566–3573.

Hammond, J. 1981 Viruses occurring in *Plantago lanceolata* in England. *Pl. Path.* **30**, 237–243.

Holland, J., Spindler, K., Horodyski, F., Grabau, E., Nichol, S. & VandePol, S. 1982 Rapid evolution of RNA genomes. *Science, Wash.* **215**, 1577–1585.

Hutson, V. & Law, R. 1981 Evolution of recombination in populations experiencing frequency-dependent selection with time delay. *Proc. R. Soc. Lond.* B **213**, 345–359.

Jaenike, J. 1978 An hypothesis to account for the maintenance of sex within populations. *Evol. Theor.* **3**, 191–194.

Kelley, S.E. 1989*a* Experimental studies of the evolutionary significance of sexual reproduction. V. A field test of the sib competition lottery hypothesis. *Evolution* **43**, 1054–1065.

Kelley, S.E. 1989*b* Experimental studies of the evolutionary significance of sexual reproduction. VI. A glasshouse test of the sib competition hypotheses. *Evolution* **43**, 1066–1074.

Kelley, S.E. 1993 Viruses and the advantage of sex in *Anthoxanthum odoratum*: a review. *Pl. Sp. Biol.* **8**, 217–223.

Kelley, S.E., Antonovics, J. & Schmitt, J. 1988 A test of the short-term advantage of sexual reproduction. *Nature, Lond.* **331**, 714–716.

Kennedy, J.S., Day, M.F. & Eastop, V.F. 1962 *A conspectus of aphids as vectors of plant viruses*. London: Commonwealth Institute of Entomology.

Kirkley, A.F. 1993 The adaptive significance of progeny dispersal in relation to viral disease. (M.S. thesis, Washington State University, Pullman.)

Kondrashov, A.S. 1984 Deleterious mutations as an evolutionary factor. 1. The advantage of recombination. *Genet. Res.* **44**, 199–217.

Kondrashov, A.S. 1988 Deleterious mutations and the evolution of sexual reproduction. *Nature, Lond.* **336**, 435–440.

Kurstak, E. (ed.) 1981 *Handbook of plant virus infections*. New York: Elsevier.

Lively, C.M. 1987 Evidence from a New Zealand snail for the maintenance of sex by parasitism. *Nature, Lond.* **328**, 519–521.

Lively, C.M., Craddock, C. & Vrijenhoek, R.C. 1990 Red Queen hypothesis supported by parasitism in sexual and clonal fish. *Nature, Lond.* **344**, 864–866.

MacClement, W.D. & Richards, M.G. 1956 Viruses in wild plants. *Can. J. Bot.* **34**, 793–799.

MacKenzie, S. 1985 Reciprocal transplantation to study local specialisation and the measurement of components of fitness. Ph.D. thesis, University College of North Wales, Bangor.

Maynard Smith, J. 1978 *The evolution of sex*. Cambridge University Press.

Michod, R.E. & Levin, B.R. 1988 *The evolution of sex: an examination of current ideas*. Sunderland, Massachusetts: Sinauer Associates.

Power, A.G. 1991 Virus spread and vector dynamics in genetically diverse plant populations. *Ecology* **72**, 233–241.

Rice, W.R. 1983 Parent-offspring pathogen transmission: a selective agent promoting sexual reproduction. *Am. Nat.* **121**, 187–203.

Rochow, W.F., Hu, J.S., Forster, R.L. & Hsu, H.T. 1987 *Pl. Dis.* **71**, 272–275.

Rochow, W.F. & Duffus, J.E. 1981 Luteoviruses and yellows diseases. In *Handbook of plant infections and comparative diagnosis* (ed. E. Kurstak), pp. 147–170. Amsterdam: Elsevier.

Schmitt, J. & Antonovics, J. 1986 Experimental studies of the evolutionary significance of sexual reproduction. IV. Effects of neighbor relatedness and aphid infestation on seedling performance. *Evolution* **40**, 837–842.

Shepherd, R.J., Richins, R.D., Duffus, J.I. & Handley, M.K. 1987 Figwort mosaic virus: properties of the virus and its adaptation to a new host. *Phytopathology* **77**, 1668–1673.

Steinhauer, D.A. & Holland, J.J. 1987 Rapid evolution of RNA viruses. *A. Rev. Microbiol.* **41**, 409–433.

Tooby, J. 1982 Pathogens, polymorphism, and the evolution of sex. *J. theor. Biol.* **97**, 557–576.

Williams, G.C. 1975 *Sex and evolution*. Princeton, New Jersey: Princeton University Press.

Yahara, T. & Oyama, K. 1993 Effects of virus infection on demographic traits of an agamospermous population of *Eupatorium chinense* (Astevaceae). *Oecologia* **96**, 310–315.

Yarwood, C.E. 1979 Host passage effects with plant viruses. *Adv. Virus Res.* **25**, 169–190.

Zimmern, D. 1988 RNA viruses. In *RNA genetics*, vol. 2 (ed. E. Domingo, J. J. Holland & P. Ahlquist), pp. 211–240. Boca Raton, Florida: CRC Press.

Discussion

J. SHYKOFF (*Swiss Federal Institute of Technology, Zurich, Switzerland*). What does Dr Kelley know about the variability of viruses in his field system? If there are different foci of foundress aphids invading the grass population bringing different viruses, then virus genotypes in the different field plots could be completely different. Genotypes specific infectivity or virulence could produce the inconsistent results he observed for the three neighbour treatments?

S. E. KELLEY. Dr Shykoff brings up an excellent point. Different viral strains and different aphid clones located in throughout the field could explain the inconsistent results. In a recent experiment, we found evidence that particular genotypes appeared to be resistant to infection by local aphid vectors/BYDV-variants, but not to infection by aphid vectors/BYDV-variants found at greater distances from where the genotypes had been sampled. This might indicate the presence of different BYDV strains locally distributed throughout the field and the subsequent buildup of local resistances. Alternatively, it might be caused by resistance to feeding by local aphid clones. Additional experiments are planned to sort this out. BYDV strains might not only be differentially infective, but might have different fitness consequences for infected hosts and this deserves further investigation as well.

4

Mate choice and maternal selection for specific parasite resistances before, during and after fertilization

CLAUS WEDEKIND

Abteilung Verhaltensökologie, Zoologisches Institut, Universität Bern, CH-3032 Hinterkappelen, Switzerland

SUMMARY

As Hamilton & Zuk pointed out, some loci may be of special importance for sexual selection because they play a crucial role in the co-evolution between parasites and hosts. In previous work I have tried to extend Hamilton & Zuk's parasite hypothesis for sexual selection, partly by including findings of immunologists and endocrinologists: in some species, handicapping signals may specifically reveal the current needs of the immune system which depends on the host's susceptibilities to different parasites. In other species, depending on the constellation of some key variables, non-handicapping signals could directly reveal the identity of resistance genes. Despite the general conflict of interests between the sexes, sexual selection may, in these cases, lead to signallers (i.e. mostly the males) focusing on improving their offspring's survival chances instead of trying to maximize their number. Males achieve this by allowing choosy females to optimize costs and benefits of each resistance.

Both parts of the extended parasite hypothesis suggest that female choice for specific heritable mate-qualities aim to optimize the resistance genetics of the unfertilized eggs. However, intersexual selection could go further than just choosing a mate. Here, I list the possible selection levels at which the mother and/or her ova could select for specific sperm haplotypes before, during and after the formation of the zygote. For many of these possible selection levels, evidence suggests that selection after mating might favour heterozygosity or even certain specific allele combinations at loci which are involved in the parasite–host co-evolution (e.g. the major histocompatibility complex or the transferrin locus).

1. INTRODUCTION

At breeding time, members of one sex normally compete more intensively for being chosen by the other sex. Mostly, the females choose (see, e.g., Clutton–Brock & Parker 1993), and many studies have shown that they use male signals as a basis. However, it still remains unsolved *why* they use particular signals. Is there information in these cues especially relevant for the female's fitness manifested in her descendants?

Hamilton & Zuk (1982) suggested that mate choice could be based on signals which reveal a male's health and vigour. Health and vigour is dependent on a male's parasite load, which itself depends on his resistance genetics. By choosing a healthy male a female tends to acquire for her offspring those resistances which are at the moment important against the predominant parasites.

This suggestion implies that there are certain loci in the host genome which are under natural *and* sexual selection. Such loci should play a crucial role in the co-evolution between parasites and hosts. Since competition between these biological systems could produce rapid changes in the genetic optima over

time, sexual selection on these loci may be a means to strongly and effectively react to an important environmental selection pressure. Moreover, this may be one of the most substantial benefits of sexual reproduction itself.

In Hamilton & Zuk's (1982) mating scenario, every male should try to signal best health and vigour to get chosen. Therefore, the signals could only be reliable if they are impossible to be cheated or if they are costly to produce. The latter argument, i.e. the 'handicap principle', has been suggested by Zahavi (1975, 1977) and proven to be feasible by Grafen (1990*a,b*) and others. However, Hamilton & Zuk's basic idea could be extended in two ways: (i) with handicapping signals revealing detailed information (Wedekind, 1992), or (ii) with cheap signals revealing the same (Wedekind 1994) (figure 1). Moreover, sexual selection on paternal resistance genes could still take place after mating, i.e. before, during and after the formation of the zygote.

Below, I will briefly outline the extended parasite hypothesis for sexual selection. Thereafter, I will propose several possible levels for maternal selection for certain allele combinations after mating, listing literature which gives some support to my suggestions.

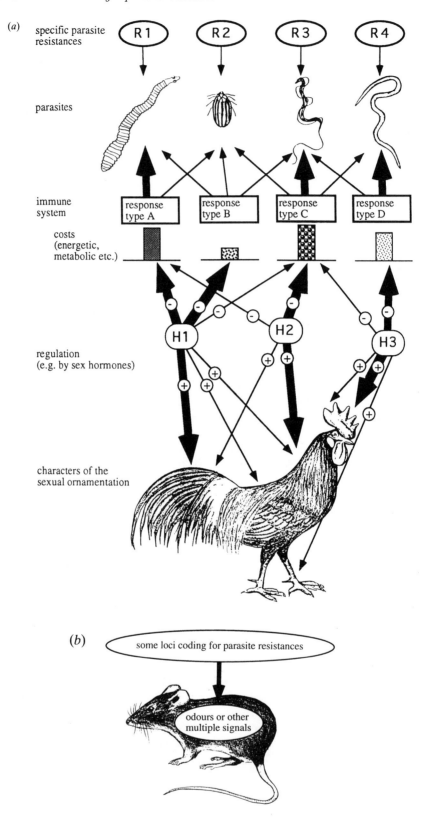

Figure 1. Schematic illustration of the extended parasite hypothesis (*a*) with the need for handicapping signals, i.e. with signals whose reliability bases on their energetic and metabolic costs to a parasitized signaller. While building the different ornaments, a signaller should optimally weigh the redistribution of his resources according to the parasites that have to be fought against, i.e. resources that are urgently needed by the immune system to fight a certain infection should be invested there while other resources could be withdrawn from the immune system for building the ornaments. Therefore, the relative and absolute extent of the different ornaments could signal detailed information about the signaller's parasites, which themselves should depend on his resistances. To optimally use this information, a female's preference should be coupled with her own resistances to achieve an optimal mix of resistance genes in the progeny. (*b*) Under some circumstances the signals need not be handicapping but can still bear detailed information about the signaller's resistance genes (see text). The connection between resistance genes and the signal could be very direct, e.g. odours as multiple signals could contain volatile parts of the gene products themselves.

2. THE EXTENDED PARASITE HYPOTHESIS

(a) *With handicapping signals*

Males of many species display several ornaments (e.g. the cock in *Gallus gallus*), or single ornaments which could vary in several properties, e.g. the red coloration of many fish often varies in colour intensity (dull to bright) as well as in colour quality (yellowish to purple-red). If these multiple ornament are costly to produce or to maintain, a male might be forced to re-allocate his resources such as energy and certain metabolites from other uses. Naturally, he should be selected to optimally weigh the redistribution of any essential resources as based on the needs of every tissue as well as the benefit expected from his signal.

The immune system might be one of the few systems whose needs may greatly differ between males, as these needs depend on the actual pathogen load that has to be fought. Therefore, the kind and amount of resources that can be withdrawn from the immune system might reflect the kind and burden of different parasites the male carries. As a consequence, detailed information about parasites could be encoded by the relative and absolute extension of the different ornaments or ornament properties.

Findings from endocrinologists support this view: steroids seem to regulate this redistribution of resources because they are known to induce the development of the ornamentation while suppressing the activity of the immune system (reviews e.g. in Grossman 1985; Folstad & Karter 1992, however, Folstad & Karter's evolutionary explanation of this phenomenon has been critically discussed in Wedekind & Folstad 1994). Every sex hormone might have a different influence on the ornamentation and on the immune system which could ensure a very precise allocation of resources. This regulation, however, need not be restricted to the action of hormones. (see figure 1*a* for a schematic overview)

The multiple ornamentation of roach (*Rutilus rutilus*, a common European fish), for example, is induced by several sex hormones (Wiley & Colette 1970), and seems to encode detailed information about the males' current parasite load (Wedekind 1992). Also, some poultry breeders seem to be able to estimate the kind of diseases of their fowl from some of the secondary sexual characters (Zuk 1984).

Females could try to use such detailed information, indeed could try to reach an optimal level of each resistance in the offspring. In the most extreme case – which is at the moment purely hypothetical – they would be able to weigh the costs and benefits of each specific resistance and combine their own and those of their mate according to the expected parasite pressure. Here, several assumptions are made whose plausibility is still to be proven: (i) the females should be able to accurately decode several male signals, (ii) they should 'know' their own level of resistances, and (iii) they should be able to estimate the future parasite pressure.

At the moment, little is known about the ability of females to be discriminating to this degree. However, some speculations exists which support points (ii) and (iii): The immune system and the nervous system seem to communicate back and forth (Blalock 1984, 1994). This has led Blalock to suggest that the immune system might act as a sensory organ for 'feeling' for pathogens which could not be recognized by other sensory systems. Furthermore, females with different resistances might show, innately or otherwise, different mate preferences, opposite to the predictions derived in the original parasite hypothesis where all females are assumed to go for the most healthy male (Hamilton & Zuk 1982). Few studies have tested this new prediction. In roach, however, different females tend to exhibit different mate preferences (Wedekind 1994). Also, studies which have demonstrated additive genetic variance in female preference could indicate the proposed polymorphism in mate preferences (see reviews in Bakker 1990; Bakker & Pomiankowski 1994), although these observations have been mainly discussed as evidence for a Fisher–Lande process (Fisher 1958; Lande 1981).

(b) *Without the need for handicapping signals*

Although the necessary prerequisite for the handicap principle, namely the conflict of interests between sender and receiver of an information, is probably always present between the sexes during mate choice (Clutton–Brock & Parker 1993), there might be conditions that allow for a liberation from the handicap principle (Wedekind 1994).

To maximize their fitness, males (or females) may sometimes do better by switching their mating strategy from trying to get as many offspring as possible to trying to enhance the survival rate of fewer offspring. They could achieve this by honestly signalling their parasite resistances because this would allow choosy females (or males, respectively) to optimize cost and benefits of their own resistances. If cheating, i.e. showing dishonestly the resistance type most often preferred, is not possible (because cheaters need to have lots of information which may not be available in time) or does not pay (because honest signalling enhances the offspring survival chances, and cheaters may not be able to outweigh the lower survival chances of offspring by having more of them), handicaps are no longer necessary for ensuring the reliability of signals.

Therefore, there could exist species or populations where cheap and unspectacular signals encode detailed information about parts of the resistance genetics of the signaller. This could be the case for many mammals since they often base their mate choice on body odours which may be cheap to produce.

3. POSSIBLE SELECTION LEVELS FOR SPECIFIC ALLELE COMBINATIONS

Figure 2 indicates the different selection levels at which sexual selection for specific allele combination has been shown or would be plausible. In the following, the discussion is concentrated on mice and humans (as these two species are best studied in this

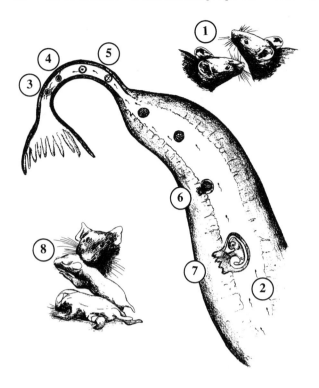

Figure 2. Schematic illustration showing the possible selection levels at which females or their ova could select for heterozygosity or even specific allele combinations in the offspring on loci which are important in the parasite–host co-evolution: (1) mate choice, (2) selection on sperm by the female within her reproductive tract, (3) egg choice for the fertilizing sperm, (4) meiotic drive influenced by the fertilizing sperm, (5) selection on the early embryo by the oviduct, (6) implantation, (7) nutrition supply to the embryo and spontaneous abortion, (8) selective feeding or selective killing of newborns.

respect), and on loci which are known to be important in the parasite–host co-evolution. I will therefore not include the *t*-complex which has been shown to influence mate choice (review in Lenington *et al.* 1992) and reproduction (for reviews see Bennett 1975; Klein 1986), but does probably not play any role in the parasite–host co-evolution.

(a) Mate choice

Behavioural ecologists who study mate choice normally measure some characteristics of possible signals of one sex and the preference for them of the other sex (reviews in Møller 1990a; Read 1990). Normally, this approach has to deal with disturbing influences of many different variables and lots of measuring errors, while large sample sizes are difficult to achieve. These may be some of the reasons why only few such studies have found evidence for a significant genetic influence on signals and/or preference (Møller 1990b; Bakker 1993). Furthermore, behavioural ecologists very often concentrate on spectacular secondary sexual characters like bright colours or long feathers. Here, the connection between genes and signals is not expected to be very direct (see section 2(a) and figure 1a).

On the other side, while studying histocompatibilities on congenic inbred mouse strains, immunologists have accidentally found strong influences of genes of the major histocompatibility complex (MHC) on mate choice (Yamazaki *et al.* 1976; Egid & Brown 1989; Potts *et al.* 1991; for a comprehensive introduction into the natural history of the MHC, see Klein 1986). Here, the signals used are not very spectacular. Information about an individual's MHC can be sniffed in body odours, urinary odours or odours of faeces, either by conspecifics in mice (Yamazaki *et al.* 1979) and rats (Singh *et al.* 1987; Brown *et al.* 1989) or by members of other mammalian species, e.g. rats on mice odours (Beauchamp *et al.* 1985), humans on mice odours (Gilbert *et al.* 1986), and mice on human odours (Ferstl *et al.* 1990). Mice acquire their MHC-dependent odour preferences at least partly during ontogenesis (Yamazaki *et al.* 1988). For recent reviews on mate choice and MHC see e.g. Boyse *et al.* (1987) and Potts & Wakeland (1993).

Shortly before Yamazaki *et al.* (1976) first found that the MHC influences mate choice, Zinkernagel & Doherty (1974) discovered that the MHC plays an important role in the immune defence against pathogens. Therefore, MHC genes are resistance genes. Accordingly, the presence of one allele or another correlates with specific parasite susceptibilities of its bearer (review in Tiwari & Terasaki 1985). MHC genes are codominant, i.e. heterozygotes can respond to any pathogens recognized by either parental MHC haplotype (Doherty & Zinkernagel 1975). Therefore, heterozygotes may often have a selective advantage.

Boyse *et al.* (1987) estimated that about half of the genetically based odour individuality in mice is determined by the MHC, while the other half may be influenced by other genes. These genes could have an impact on mate choice as well. Suspicious candidates might be loci which are polymorphic and have one or several functions in the host's defence against pathogens, e.g. the transferrin loci (see below). To my knowledge, however, any influences of these loci on mate choice have not been studied so far.

In most species where mate choice has been studied, a chosen male is diploid and often heterozygous at the interesting loci. Therefore, females who choose a mate for his alleles often need further selection mechanisms to ensure heterozygosity at the important loci or to reach epistatic allele combinations with a selective advantage.

(b) Haplotype-specific signals on sperm membrane and survival of sperm until the oviduct

The female reproductive tract is very hostile to sperm. Physiochemical and immunological factors in the vagina and the cervix play an important role in sperm survival and transport (e.g. Overstreet 1983; Hafez 1987; Birkhead *et al.* 1993). In many mammals, most of the sperm of an ejaculate do not even pass the cervix (Mortimer 1983; Overstreet 1983). Until reaching the oviduct the sperm have to survive strong selection by physical and chemical barriers, phagocytosis by leukocytes and high concentrations of antisperm antibodies which coat a majority of ejaculated sperm (Overstreet 1983; Birkhead & Møller 1993; Birkhead *et al.* 1993). Therefore, maternal selection for specific spermal haplotypes could be possible by several means between the vagina and the oviduct. However, as a necessary prerequisite, sperm should signal their haplotype with respect to resistance alleles. Therefore, haplotype-specific cell surface components should be transcribed and produced after meiosis.

In general, haploid expression of mammalian genes in sperm is possible, because during spermatogenesis, e.g. in mice, RNA synthesis is terminated only during the early postmeiotic phase (Eddy *et al.* 1993). Protein synthesis continues until late in the postmeiotic phase (Eddy *et al.* 1993).

Some of the surface antigens on spermatozoa seem to be transcribed from the haploid genome. Several studies using different methods have demonstrated that at least the MHC and the *t*-complex of the haploid genome may be expressed in the sperm membrane (haploid expression of the MHC-type in humans: class I: Fellous & Dausset 1970; Arnaiz-Villena & Felsenstein 1976; Halim *et al.* 1982; class II: Halim & Felsenstein 1975; haploid expression of the *t*-complex in mice: Yanagisawa *et al.* 1974; haploid signalling of the MHC in mice suggested by Goldberg *et al.* 1970). However, these findings may still be controversial, because Haas & Nahhas (1986) could not find any MHC antigens on human sperm, and Kurpisz *et al.* (1987) reported that sperm of only few men in their sample showed HLA expression.

Nicol & McLaren (1974) found a strong influence of female genotype on sperm transport in two mouse strains. Furthermore, about 1.5 h after mating females of both strains tended to have more sperm in their oviduct if mated with males of the other strain each than if mated with males of their own strain. These findings suggests that the female reproductive system may favour certain sperm genotypes according to their own genotype. Since the sperm probably signal specifically their own MHC-haplotype, the females may select for certain MHC-types in their reproductive tract. Since MHC-heterozygotes may well have higher fitness expectations (see previous chapter), maternal choice for MHC-haplotype in sperm could be adaptive.

Further support for this hypothesis comes from the self-nonself recognition system which is known in the angiosperms. The 'self-incompatibility' (SI) is a genetically controlled mechanism which ensures that a plant is fertilized by a genetically dissimilar individual of the same species (Haring *et al.* 1990). If the pollen that landed on the stigma carries the same allele as one of the two in the pistil, the female tissue will hinder the pollen from growing into the pistil.

In vertebrates there is some evidence that the MHC is not the only system which could be important in parasite-driven maternal selection: Transferrin is a protein that binds iron and zinc. In most vertebrates studied this protein shows a polymorphism within a population (Ashton & Dennis 1971). In humans,

transferrin is encoded by a single locus on chromosome 3 with codominant expression. The protein has several important functions in immunoregulation (Vodinelich *et al.* 1983) and in defence against pathogens (Weinberg 1978). Since it also shows a polymorphism it seems to be one of the important genes involved in host–parasite co-evolution.

Canham *et al.* (1970) found anomalous segregation at the transferrin locus of the deer mouse (*Peromyscus maniculatus borealis*) in a breeding experiment. The same has been found in mice and other species (Ashton & Dennis 1971, and references therein). Canham *et al.* (1970) and Ashton & Dennis (1971) listed several arguments supporting their favoured hypothesis that selective sperm transport might be the cause for the observed deviations in phenotype frequencies. In humans, Weitkamp *et al.* (1985) found that the maternal transferrin genotype influences the transmission ratio of the transferrin genes from heterozygous fathers to the offspring. They discussed this distortion as probable maternal selection on sperm during their way to the oviduct, at the egg itself (see next section), or even as maternal selection on the early embryo (see section 3(f)). Hence, it is still not clear where the selection on paternal transferrin type really takes place. However, as transferrin lies on a different chromosome from the MHC it is evident that it is probably another important gene in a parasite-driven maternal selection in vertebrate (next to MHC).

(c) Egg choice for sperm

Normally, oocytes are surrounded by envelopes such as the zona pellucida in mammals or the vitelline envelopes in amphibians or many invertebrates. Both the zona pellucida and the vitelline membrane appear to contain specific receptors for the binding of the spermatozoa (Moore & Bedford 1983; Bazer *et al.* 1987). Here, the possibility for sperm selection by the ovum is evident.

The mammalian zona pellucida is important in the initial stages of fertilization: it plays a role in blocking polyspermy, and it is, in some species, important in sperm capacitation, i.e. the changes that enable the sperm to fertilize the egg (Dunbar 1983). Furthermore, the sperm must bind to the zona and penetrate it before fusing with the oocyte plasma membrane. Several zona antigens are important in the sperm–egg interactions. However, the communication between sperm and egg which takes place at the zona pellucida and, on a later stage, at the egg membrane itself is not yet fully understood (review in Dunbar 1983; Moore & Bedford 1983).

Gametes of many lower organisms (yeast, algae, fungi and protozoa) seem to choose mates on the basis of pheromones (Pagel 1993 and references therein). In higher organisms, however, eggs may even be able to choose for a specific sperm haplotype. *Botryllus* sp., for example, are colonial tunicates which are likely to be sometimes subject to natural tissue transplantations. The *Botryllus* fusibility is controlled by a polymorphic locus with similar properties to the vertebrate MHC.

Scofield *et al.* (1982) found that this locus not only controls allorecognition but also the fusion of gametes: *Botryllus* eggs resisted fertilization by sperm from the same colony for longer time than by sperm from a strange allele on the fusibility locus. Thus, the eggs strongly select for heterozygosity on this locus in the zygote.

(d) Meiotic drive influenced by sperm haplotype

The eggs of most vertebrate and invertebrate species are in some stage of meiotic arrest at the time of fertilization (exceptions are the coelenterates and echinoids, which complete meiosis prior to ovulation (Lopo 1983)). In mammals, the second maturational division is completed only when the sperm has digested a pathway through the zona pellucida and is penetrating the vitelline membrane of the ovum. Then, the first and/or the second polar body is expelled into the perivitelline space (Wolgemuth 1983; Bazer *et al.* 1987; Hafez 1987).

The significance of this suspension of meiosis is not yet clear but in the context of this paper it raises a suspicion. The decision of the ovum (or parental zygote) about which haplotype will go into the zygote and which will be lost in the polar body could be influenced by the haplotype of the fertilizing sperm to ensure higher fitness of the zygote. But again, several prerequisites need to be fulfilled to make this plausible: (i) the ovum should be able to detect the sperm's haplotype on the important loci, and (ii) it should have a regulatory mechanism to decide accordingly which of its own haplotypes will stay in the zygote.

After fusion of the membranes of the two gametes the spermatozoan nuclear envelope disintegrates and the released chromatin material undergoes decondensation for a short time (Wolgemuth 1983). This decondensation seems to be at least partly controlled by the egg since it requires specific components (e.g. the 'sperm nucleus-decondensing factor') in the ovum's cytoplasm (Wolgemuth 1983; Bazer *et al.* 1987). Here, the egg could have the possibility of analysing the allelic specificity of the spermal haplotype, as an alternative to the potential signalling between the gametes before fusion (see above). Correspondingly, it appears that transcription of pronuclei occurs, i.e. before fusion of the paternal and maternal genomes and probably even before the egg has decided which of the two maternal haplotypes goes into the polar body (Wolgemuth 1983). After decondensating, the sperm nuclear envelope is replaced by a new envelope provided by the egg's cytoplasm before male and female pronuclei migrate to the ovum centre (Bazer *et al.* 1987).

(e) Zygotic cleavage and implantation

If it is beneficial for the mother to be able to select after fertilization she should decide as soon as possible whether to support an embryo or not, minimizing her costs. However, the formation of the zygote is the start of the conflict between the embryo and the mother

(Haig 1993), i.e. the embryo might develop some mechanism of self-protection.

The oviductal epithelium is in close contact to the early embryo and most active near it. Therefore, oviductal secretions seem to provide metabolic support for the early embryo (Hafez 1987). Also, the descent of the embryo depends strongly on the oviduct. A premature entry of the embryo into the uterus (e.g. during the morula stage) often causes its degeneration (Hafez 1987). After entrance of the blastocyst into the uterus, it must implant into the uterus tissue to establish the necessary close contact to the mother during pregnancy. Until this stage, a majority of embryos get lost for various reasons (e.g. for humans see Chen 1986). It is still unclear whether maternal selection for certain offspring genotypes is one of these reasons. However, several possibilities for maternal selection might be given from the very beginning of the embryo's existence.

Until reaching the uterus the embryo is still covered by the zona pellucida (Bazer *et al.* 1987). This could be a mechanism of self-protection of the early embryo against the mother, since for example, the sperm's MHC type can be identified in the membrane of the early embryo as late as the eight-cell stage (Bazer *et al.* 1987) while neither MHC class I nor class II molecules are detectable on the zona pellucida. Discussing the case in humans, Desoye *et al.* (1988) could not find any MHC molecules on the blastocysts which are released from the zona pellucida and are implanted into the uterus tissue.

(f) Embryo growth, spontaneous abortion or resorption

Maternal selection could also take place at a later stage during the embryo's growing, still selecting for certain offspring genotypes on loci which have a selective advantage in the host–parasite co-evolution. In humans 10–25% of clinically recognized pregnancies end in spontaneous abortion (Chen 1986). Some of these abortions seem to be caused by immunological factors. In many human populations it has recently been shown that couples who suffer from recurrent spontaneous abortion (RSA) share a higher proportion of MHC fragments compared to control families. For the U.S.A. this is discussed by Beer *et al.* (1985), for Germany by Karl *et al.* (1989), for a Chinese population in Taiwan by Ho *et al.* (1990), for Japan by Koyama *et al.* (1991) and for Finland by Laitinen (1993). Further examples are given in these studies.

Reznikoff-Etievant *et al.* (1991) found in a French population that newborn children of RSA-couples with a high incidence of MHC-sharing showed a significant lower birth weight. Correspondingly, Billington (1964) had found in mice that heterozygote foetuses have larger placentae and are considerably larger themselves than homozygous ones in mothers of two different strains. The birth weight and the size of the placenta could reveal a mother's 'decision' about her investment into the embryo. This view is supported by further findings on mice, hamsters and rats (Beer *et al.* 1975). (It should be stressed that this decision is most probably no more a conscious one than is the decision to reject an allograft.) However, there are of course other possible explanations. It should be recalled that heterozygosity at the MHC is likely to correlate with heterozygosity at many other loci. It may simply be heterosis expressed at some of these (as well as, possibly, at MHC) that causes a general advantage independent of their mother's decision.

Spontaneous abortions are better understood in mice. Bruce (1959) discovered that the blocking of pregnancy can be induced by the odour of a strange male. This so-called 'Bruce effect' strongly depends on the MHC of the second male which induces the pregnancy block (Yamazaki *et al.* 1983). The incidence of the pregnancy blocking was higher if the MHC of the second male differed from the one of the father of the embryos (the authors used congenic inbred strains which only differed in their MHC type). Therefore, the pregnancy block in mice may sometimes reveal a strategic decision of the mother which is at least partly influenced by the MHC.

For an evolutionary explanation of the observed MHC-sharing in RSA couples, Verrell & McCabe (1990) suggested that the MHC could be a marker for relatedness in inbred populations, indicating the possible sharing of deleterious alleles in related mates. Females may abort in order to prevent too much investment in inbred offspring of low reproductive value. They stressed, however, that their suggestion could only explain RSAs in highly inbred populations.

In agreement with my suggested evolutionary explanation, the MHC does not seem to be the only genetic system which is both important in pathogen defence and in embryo growth (compare section 3(b)). Already Weitkamp *et al.* (1985) observed an increased frequency of transferrin *C3* in couples with recurrent spontaneous abortions. Although they did not investigate the possible role of the foetal transferrin genotype, they hypothesized that different combinations of maternal and paternal transferrin genes could affect the embryonic development differently. Correspondingly, a dense concentration of transferrin and transferrin receptors have been found at the human materno–foetal interface (Faulk & Galbraith 1979).

(g) Selective killing or neglect of newborns

Poor parental care and maternal cannibalism are well documented and common in many rodent species (Elwood 1992). Often, maternal cannibalism could be of adaptive value. By killing her weakest offspring the mother might stop investments in young that are sick and therefore unlikely to survive. Another possibility is allocation of resources to young in a selective manner. In poor environmental conditions mothers often kill and consume the smallest pups to minimize wasted investment and to provide their larger pups with sufficient milk. This is known for mice, but is even more common in hamsters, where females typically produce more pups than they can rear. Filial cannibalism has also been documented in birds of 13 different families (Stanback & Koenig 1992). In these species, too, parents normally kill the weakest chicks.

4. CONCLUSIONS AND PERSPECTIVES

One of the most substantial benefits of sexual reproduction itself could be that it allows animals to rapidly react to a continuously changing environmental selection pressure, e.g. co-evolving parasites ('Red Queen Hypothesis', e.g. Ladle 1992). This counteraction would be most efficient if the females (or their ova) were able to encourage for allele combinations in the progeny either by direct and specific favour or by favouring manifest vigour. Such selection should mainly act on loci which are crucial in the arms race in co-evolving systems, e.g. polymorphic loci which encode for parasite resistances in the host. Good candidates for which some evidence already exists may be the MHC and the transferrin locus, but many other polymorphic defense-systems loci may be implicated (Hamilton *et al.* 1990; Hamilton 1993).

Sexual selection for parasite resistances need not depend on display and the handicap principle. Under certain circumstances, cheap signals showing detailed information about the identity of resistances could be reliable. Furthermore, selection on these loci could take place at many different levels and times, both before mating ('mate choice'), and afterwards, with the afterwards itself including events before, during or after the formation of the zygote ('maternal selection'). Regarding some of the possible selection levels I have listed there is already suggestive evidence that maternal selection takes place on the MHC and the transferrin locus. However, much effort is still necessary to understand the described phenomena. Several important question are largely unsolved, e.g. whether females just select for heterozygosity on certain loci in the offspring or whether they actively seek allele combinations which might be beneficial under the current conditions. To progress it is necessary to understand better the specific allele combinations advantageous under given environmental conditions. Such understanding will clarify whether, for example, MHC-induced mate preferences in mice are simply to avoid inbreeding or are better treated as part of the rubric of good-genes sexual selection (see, e.g., the discussion in Potts & Wakeland 1993).

I very much thank Manfred Milinski, William D. Hamilton, Thomas Rülicke, Michel Chapuisat, Markus Frischknecht, Erwin Macas, Laurent Keller and Beda Stadler for stimulating discussions and comments on the manuscript.

REFERENCES

Arnaiz–Villena, A. & Festenstein, H. 1976 HLA genotyping by using spermatozoa: evidence for haploid gene expression. *Lancet* 707–709.

Ashton, G.C. & Dennis, M.N. 1971 Selection at the transferrin locus in mice. *Genetics* **67**, 253–265.

Bakker T.C.M. 1990 Genetic variation in female mating preference. *Neth. J. Zool.* **40**, 617–642.

Bakker, T.C.M. 1993 Positive genetic correlation between female preference and preferred male ornament in sticklebacks. *Nature, Lond.* **363**, 255–257.

Bazer, F.W., Geisert, R.D. & Zavy, M.T. 1987 Fertilization, cleavage and implantation. In: *Reproduction in farm animals* (ed. E. S. E. Hafez), pp. 210–228. Philadelphia: Lea & Febiger.

Beauchamp, G.K., Yamazaki, K., Wysocki, C.J., Slotnick, B.M., Thomas, L. & Boyse, E.A. 1985 Chemosensory recognition of mouse major histocompatibility types by another species. *Proc. natn. Acad. Sci. U.S.A.* **82**, 4186–4188.

Beer, A.E., Scott, J.R. & Billingham, R.E 1975 Histocompatibility and maternal immunological status as determinants of fetoplacental weight and litter size in rodents. *J. exp. Med.* **142**, 180–196.

Beer, A.E., Semprini, A.E., Zhu, X.Y. & Quebbeman, J.F. 1985 Pregnancy outcome in human couples with recurrent spontaneous abortions: HLA antigen profiles, HLA antigen sharing, female serum MLR blocking factors, and paternal leukocyte immunization. *Exp. Clin. Immunogenet.* **2**, 137–153.

Bennett, D. 1975 The *t*-locus of the mouse. *Cell* **6**, 441–454.

Billington, W.D. 1964 Influence of immunological dissimilarity of mother and foetus on size of placenta in mice. *Nature, Lond.* **202**, 317–318.

Birkhead, T.R. & Møller, A.P. 1993 Female control of paternity. *Trends Ecol. Evol.* **8**, 100–104.

Birkhead, T.R., Møller, A.P. & Sutherland, W.J. 1993 Why do females make it so difficult for males to fertilize their eggs? *J. theor. Biol.* **161**, 51–60.

Blalock, J.E. 1984 The immune system as a sensory organ. *J. Immunol.* **132**, 1067–1070.

Blalock, J.E. 1994 The immune system our sixth sense. *Immunologist* **2**, 8–15.

Boyse, E.A., Beauchamp, G.K. & Yamazaki, K. 1987 The genetics of body sent. *TIG* **3**, 97–102.

Brown, R.E., Roser, B. & Singh, P.B. 1989 Class I and class II regions of the major histocompatability complex both contribute to individual odors in congenic inbred strains of rats. *Behav. Genet.* **19**, 659–674.

Bruce, H.M. 1959 An exteroceptive block to pregnancy in the mouse. *Nature, Lond.* **184**, 105.

Canham, R.P., Birdsall, D.A. & Cameron, D.G. 1970 Disturbed segregation at the transferrin locus of the deer mouse. *Microbiol. Rev.* **16**, 355–357.

Chen, C. 1986 Early reproductive loss. *Aust. N.Z. J. Obstet. Gynaecol.* **26**, 215–218.

Clutton-Brock, T.H. & Parker, G.A. 1993 Potential reproductive rates and the operation of sexual selection. *Q. Rev. Biol.* **67**, 437–456.

Desoye, G., Dohr, G.A., Motter, W. *et al.* 1988 Lack of HLA class I and class II antigens on human preimplantation embryos. *J. Immunol.* **140**, 4157–4159.

Doherty, P.C. & Zinkernagel, R.M. 1975 Enhanced immunological surveillance in mice heterozygous at the *H-2* gene complex *Nature, Lond.* **256**, 50–52.

Dunbar, B.S. 1983 Morphological, biochemical, and immunochemical characterization of the mammalian zonae pellucida. In: *Mechanism and control of animal fertilization* (ed. J. F. Hartmann), pp. 139–175. New York, London: Academic Press.

Eddy, E.M., Welch, J.E. & O'Brien, D.A. 1993 Gene expression during spermatogenesis. In: *Molecular biology of the male reproductive system* (ed. D. de Kretser), pp. 181–232. New York, London: Academic Press.

Egid, K. & Brown, J.L. 1989 The major histocompatibility complex and female mating preferences in mice. *Anim. Behav.* **38**, 548–550.

Elwood, R. 1992 Pup-cannibalism in rodents: causes and consequences. In: *Cannibalism. Ecology and evolution among diverse taxa* (ed. M. A. Elgar & B. J. Crespi), pp. 299–322. Oxford: Oxford University Press.

Faulk, W.P. & Galbraith, M.P. 1979 Trophoblast transferrin and transferrin receptors in the host-parasite relationship of human pregnancy. *Proc. R. Soc. Lond. B* **204**, 83–97.

Fellous, M. & Dausset, J. 1970 Probable haploid expression of HL-A antigens on human spermatozoon. *Nature, Lond.* **225**, 191–193.

Ferstl, R. Pause, B., Schüler, M. *et al.* 1990 Immune system signalling to the brain: MHC-specific odors in human. In: *Psychiatry: a world perspective (volume 2)* (ed. C. N. Stefanis *et al.*), pp. 751–755. London: Elsevier Science Publishers.

Fisher, R.A. 1958 *The genetical theory of natural selection (2nd ed.)*. New York: Dover Publications.

Folstad, I. & Karter, A.K. 1992 Parasites, bright males and the immunocompetence handicap. *Am. Nat.* **139**, 603–622.

Gilbert, A.N., Yamazaki, K., Beauchamp, G.K. & Thomas, L. 1986 Olfactory discrimination of mouse strains (*Mus musculus*) and major histocompatibility types by humans (*Homo sapiens*). *J. comp. Psych.* **100**, 262–265.

Giphart, M.J. & D'Amaro, J. 1983 HLA and reproduction? *J. Immunogenet.* **10**, 25–29.

Goldberg, E.H., Aoki, T., Boyse, E.A. & Bennett, D. 1970 Detection of *H-2* antigens on mouse spermatozoa by the cytotoxicity test *Nature, Lond.* **228**, 570–572.

Grafen A. 1990*a* Sexual selection unhandicapped by the Fisher process. *J. theor. Biol.* **144**, 473–516.

Grafen A. 1990*b* Biological signals as handicaps. *J. theor. Biol.* **144**, 517–546.

Grossman, C.J. 1985 Interactions between the gonadal steroids and the immune system. *Science* **227**, 257–261.

Haas, G.G. Jr. & Nahhas, F. 1986 Failure to identify HLA ABC and Dr antigens on human sperm. *Am. J. Reprod. Immunol. Microbiol.* **10**, 39–46.

Hafez, E.S.E. 1987 Transport and survival of gametes. In: *Reproduction in farm animals* (ed. E. S. E. Hafez), pp. 168–188. Philadelphia: Lea & Febiger.

Haig, D. 1993 Genetic conflicts in human pregnancy. *Q. Rev. Biol.* **68**, 495–532.

Halim, K. & Festenstein, H. 1975 HLA-D on sperm is haploid, enabling use of sperm for HLA-D typing. *Lancet* 1255–1256.

Halim, K., Wong, D.M. & Mittal, K.K. 1982 The HLA typing of human spermatozoa by two color fluorescence. *Tissue Antigens* **19**, 90–91.

Hamilton, W.D. 1993 Haploid dynamic polymorphism in a host with matching parasites: effects of mutation/subdivision, linkage, and patterns of selection. *J. Hered.* **84**, 328–338.

Hamilton, W.D. & Zuk, M. 1982 Heritable true fitness and bright birds: a role for parasites? *Science* **218**, 384–387.

Hamilton, W.D., Axelrod, R. & Tanese, R. 1990 Sexual reproduction as an adaptation to resist parasites (a review). *Proc. natn. Acad. Sci. U.S.A.* **87**, 3566–3573.

Haring, V., Gray, J.E., McClure, B.A., Anderson, M.A. & Clarke, A.E. 1990 Self-incompatibility: a self-recognition system in plants. *Science* **250**, 937–941.

Ho, H.N., Gill, T.J., Nsieh, R.P., Hsieh, H.J. & Lee, T.Y. 1990 Sharing of human leukocyte antigens in primary and secondary recurrent spontaneous abortions. *Am. J. Obstet. Gynec.* **163**, 178–188.

Karl, A., Metzner, G., Seewald, H.J., Karl, M., Born, U. & Tilch, G. 1989 HLA compatibility and susceptibility to habitual abortion. Results of histocompatibility testing of couples with frequent miscarriages. *Allerg. Immunol. (Leipz)* **35**, 133–140.

Klein, J. 1986 *Natural history of the major histocompatibility complex*. New York: John Wiley & Sons.

Koyama, M., Saji, F., Takahashi, S. *et al.* 1991 Probabilistic assessment of the HLA sharing of recurrent spontaneous abortion couples in the Japanese population. *Tissue Antigens* **37**, 211–217.

Kurpisz, M., Fernandez, N., Witt, M., Kowalik, I., Szymczynski, G.A. & Festenstein, H. 1987 HLA expression on human germinal cells. *J. Immunogenet.* **14**, 23–32.

Ladle, R. J. 1992 Parasites and sex: catching the Red Queen. *Trends Ecol. Evol.* **7**, 405–408.

Laitinen, T. 1993 A set of MHC haplotypes found among Finnish couples suffering from recurrent spontaneous abortions. *Am. J. Reprod. Immunol.* **29**, 148–154.

Lande, M. 1981 Models of speciation by sexual selection on polygenic traits. *Proc. natn. Acad. Sci. U.S.A.* **78**, 3721–3725.

Lenington, S., Coopersmith, C.B. & Williams, J. 1992 Genetic basis of mating preferences in wild house mice. *Am. Zool.* **32**, 40–47.

Lopo, A.C. 1983 Sperm-egg interaction in invertebrates. In: *Mechanism and control of animal fertilization* (ed. J. F. Hartmann), pp. 269–324. New York, London: Academic Press.

Møller, A.P. 1990a Parasites and sexual selection: Current status of the Hamilton and Zuk hypothesis. *J. evol. Biol.* **3**, 319–328.

Møller, A.P. 1990b Effect of a haematophagous mite on the barn swallow (*Hirundo rustica*): a test of the Hamilton and Zuk Hypothesis. *Evolution* **44**, 771–784.

Moore, H.D.M. & Bedford, J.M. 1983 The interaction of mammalian gametes in the female. In: *Mechanism and control of animal fertilization* (ed. J. F. Hartmann), pp. 453–497. New York, London: Academic Press.

Mortimer, D. 1983 Sperm transport in the human female reproductive tract. In: *Oxford reviews of reproductive biology* (ed. C. A. Finn), pp. 30–61. Oxford: Clarendon Press.

Nicol, A. & McLaren, A. 1974 An effect of the female genotype on sperm transport in mice. *J. Reprod. Fert.* **39**, 421–424.

Nordlander, C., Hammarström, L., Lindblom, B. & Smith, C.I.E.E. 1983 No role of HLA in mate selection. *Immunogenetics* **18**, 429–431.

Overstreet, J.W. 1983 Transport of gametes in the reproductive tract of the female mammal. In: *Mechanism and control of animal fertilization* (ed. J. F. Hartmann), pp. 499–543. New York, London: Academic Press.

Pagel, M. 1993 Honest signalling among gametes. *Nature, Lond.* **363**, 539–541.

Potts, W.K., Manning, C.J. & Wakeland, E.K. 1991 Mating patterns in seminatural populations of mice influenced by MHC genotyp. *Nature, Lond.* **352**, 619–621.

Potts, W.K. & Wakeland, E.K. 1993 Evolution of mhc genetic diversity: a tale of incest, pestilence and sexual preference. *TIG* **9**, 408–412.

Read, A.F. 1990 Parasites and the evolution of host behaviour. In: *Parasitism and host behaviour* (ed. C. J. Barnard & J. M. Behnke), pp. 117–157. London: Taylor and Francis.

Reznikoff Etievant, M.F., Bonneau, J.C., Alcalay, D. *et al.* 1991 HLA antigen-sharing in couples with repeated spontaneous abortions and the birthweight of babies in successful pregnancies. *Am. J. Reprod. Immunol.* **25**, 25–27.

Rosenberg, L.T., Cooperman, D. & Payne, R. 1983 HLA and mate selection. *Immunogenetics* **17**, 89–93.

Scofield, V.L., Schlumpberger, J.M, West, L.A. & Weissman, I.L. 1982 Protochordate allorecognition is controlled by a MHC-like gene system. *Nature, Lond.* **295**, 499–502.

Singh, P.B., Brown, R.E. & Roser, B. 1987 MHC antigens in urine as olfactory recognition cues. *Nature, Lond.* **327**, 161–164.

Stanback, M.T. & Koenig, W.D. 1992 Cannibalism in birds. In: *Cannibalism. Ecology and evolution among diverse taxa* (ed. M. A. Elgar & B. J. Crespi), pp. 277–298. Oxford: Oxford University Press.

Tiwari J.L. & Terasaki, P.I. 1985 *HLA and disease associations*. New York, Berlin: Springer-Verlag.

Verrell, P.A. & McCabe, N.R. 1990 Major histocompatibility antigens and spontaneous abortion: an evolutionary perspective. *Med. Hypoth.* **32**, 235–238.

Vodinelich, L., Sutherland, R., Schneider, C., Newman, R. & Greaves, M. 1983 Receptor for transferrin may be a 'target' structure for natural killer cells. *Proc. natn. Acad. Sci. U.S.A.* **80**, 835–839.

Wedekind, C. 1992 Detailed information about parasites revealed by sexual ornamentation. *Proc. R. Soc. Lond. B* **247**, 169–174.

Wedekind, C. 1994 Handicaps not obligatory in sexual selection for resistance genes *J. theor. Biol.* (Submitted.)

Wedekind, C. & Folstad, I. 1994 Adaptive or non-adaptive immunosuppression by sex hormones? *Am. Nat.* **143**, 936–938.

Weinberg, E.D. 1978 Iron and infection. *Microbiol. Rev.* **42**, 45–66.

Weitkamp, L.R. & Schacter, B.Z. 1985 Transferrin and HLA: spontaneous abortion, neural tube defects, and natural selection. *New Engl. J. Med.* **313**, 925–932.

Wiley, M.L. & Colette, B.B. 1970 Breeding tubercles and contact organ in fishes: their occurrence, structure, and significance. *Bull. Am. Mus. nat. Hist.* **143**, 145–216.

Wolgemuth, D.J. 1983 Synthetic activities of the mammalian early embryo: molecular and genetic alterations following fertilization. In: *Mechanism and control of animal fertilization* (ed. J. F. Hartmann), pp. 415–452. New York, London: Academic Press.

Yamazaki, K., Boyse, E.A., Miké, V. *et al.* 1976 Control of mating preference in mice by genes in the major histocompatibility complex. *J. exp. Med.* **144**, 1324–1335.

Yamazaki, K., Yamaguchi, M., Baranoski, L., Bard, J., Boyse, E.A. & Thomas, L. 1979 Recognition among mice. Evidence from the use of a Y-maze differentially scented by congenic mice of different major histocompatibility types. *J. exp. Med.* **150**, 755–760.

Yamazaki, K., Beauchamp, G.K., Wysocki, C.J., Bard, J., Thomas, L. & Boyse, E.A. 1983 Recognition of *H-2* types in relation to the blocking of pregnancy in mice. *Science* **221**, 186–188.

Yamazaki, K., Beauchamp, G.K., Kupniewski, D., Bard, J., Thomas, L. & Boyse, E.A., 1988 Familial imprinting determines *H-2* selective mating preferences. *Science* **240**, 1331–1332.

Yanagisawa, K., Pollard, D.R., Bennett, D., Dunn, L. C. & Boyse, E.A. 1974 Transmission ration distortion at the *T*-locus, identification of two antigenically different populations of sperm in *t*-heterozygotes. *Immunogenetics* **1**, 91–96.

Zahavi, A. 1975 Mate selection – a selection for a handicap. *J. theor. Biol.* **53**, 205–214.

Zahavi, A. 1977 The costs of honesty (further remarks on the handicap principle). *J. theor. Biol.* **67**, 603–605.

Zinkernagel, R.M. & Doherty, P.C. 1974 Immunological surveillance against altered self components by sensitised *T* lymphocytes in lymphocytic choriomeningitis. *Nature, Lond.* **251**, 547–548.

Zuk, M. 1984 A charming resistance to parasites. *Natur. Hist.* **4**, 28–34.

5

Infection and colony variability in social insects

PAUL SCHMID-HEMPEL

ETH Zürich, Experimental Ecology, ETH-Zentrum NW, CH-8092 Zürich, Switzerland

SUMMARY

The average relatedness among colony members in the social insects, such as bees, wasps and ants, is often low, contrary to the expectations of kin selection theory. Lower relatedness results from multiple mating by the queens (polyandry) or from the presence of more than one functional queen (polygyny). Among the proposed advantages for such mating systems, selection by parasites for within-colony genetic variability is discussed. Empirical studies of this problem are few, but several lines of evidence suggest a role for parasites, such that genetic diversity reduces the rate of within-group transmission. Theoretical considerations indicate that multiple mating is advantageous under conditions of low mating costs relative to parasite pressure and when intermediately sized colonies have a disproportionately large share of reproductive success in the population. In this view, mating strategies (as in polyandry) and strategies of female associations (as in polygyny) that lead to an increased genetic diversity among offspring are, at least in part, an instance of variance reduction in relation to parasitism.

1. PARASITES, SOCIAL LIFE AND MATING SYSTEMS

Parasites can be contracted and subsequently threaten many individuals in socially living species. Therefore, Alexander (1974) and Freeland (1976) suggested that seemingly unrelated phenomena in social animals, such as group aggression towards new members, the maintenance of territories, or troop size in primates, may all have evolved to avoid novel infections. For directly transmitted diseases, positive correlations between group size, number of parasite species per host and infection intensity have been demonstrated in several animal species (see, for example, Davies *et al.* 1991); negative correlations are found in vector-transmitted diseases (Poulin & Fitzgerald 1989). Once acquired, disease can be transmitted more easily within than between groups, as suggested for human smallpox in African households (Becker & Angulo 1981). In several cases, fitness effects of parasites for group-living animals have been demonstrated (see, for example, Brown & Brown 1986). This reasoning clearly demonstrates the role of parasites as selective agents in social species.

Several hypotheses have been formulated that connect parasitism with mating systems. For example, polyandry (females mating with several males) produces increased genetic diversity among offspring. Polygyny (males mating with several females) is also common and produces similar results. Particular benefits may accrue when mating is selective, reflecting the tendency of females to garner resistance alleles for offspring from high-quality males (Hamilton & Zuk 1982). Interestingly, the degree of polygyny and exposure to severe pathogens correlates positively across a number of human societies (Low 1988). Social life and mating systems are particularly intertwined in social insects: polyandry and 'polygyny' (here meaning the presence of more than one queen per colony) correlate with the evolution of sociality and the potential for cooperation or conflict. From the perspective of the parasite the primary importance of sociality is that it aggregates individuals both in space and by genotype, increasing the chance of transmission and successful infection and establishment in a new host.

2. PARASITES AND VARIABILITY IN SOCIAL INSECTS

(a) Social organization

Social insects – essentially the social Hymenoptera (ants, bees, wasps) and the termites – are a large and diverse group. In the social Hymenoptera, mother queen(s) and their daughters (the workers) live together in colonies and cooperatively raise young that become either new (sterile) workers or sexual individuals. The organization of work in social insect colonies, based on different degrees of division of labour and polymorphism, affects parasitism and is in turn affected by parasites in ways yet to be explored (Schmid-Hempel & Schmid-Hempel 1993). For example, acute bee paralysis virus in honey bees is transmitted by trophallaxis (when workers transfer liquid food to one another) (Bailey & Gibbs 1964). When the colony increases foraging activity, trophallaxis is reduced and the disease dies out (Bailey *et al.* 1983). On the other hand, infection by the microsporidian parasite *Nosema apis* causes workers of the honey bee to start foraging earlier in life and thus alters the entire work profile of the colony (Wang & Moeller 1970). In the fire ant *Solenopsis*, parasitic flies (Phoridae) attack the majors (the large worker caste) which are costly to produce (Feener 1987, 1988). Feener thus suggested that such parasitism may select against caste differentiation. Workers infected by a fungal pathogen in termites (Kramm *et al.* 1982) or by a virus in the honey bee (Waddington & Rothenbuhler 1976) are more intensively groomed by their nestmates, which are thus more likely to become infected. This is suggestive of parasites being able to manipulate secretions and/or the social behaviour of their host to their own advantage. Finally, infected workers could threaten their colony by transmitting a disease to their queen. This should be particularly relevant in species where the colony depends on a single queen, as in the leaf-cutting ants. Interestingly, in all species of social insect, it is the young workers that attend the queen while the older ones forage (Wilson 1971). As the probability of being infected by a disease must increase with age and novel diseases can be picked up by the foragers, young workers should be less likely to be an infection risk. Thus, parasitism provides additional hypotheses for the adaptive value of the standard age-dependent division of labour (Schmid-Hempel & Schmid-Hempel 1993).

Transmission to larvae is necessary to complete the life cycle for some parasites of social insects. The microsporidian *Burenella* in fire ants has to infect the fourth instar larvae (Jouvenaz *et al.* 1981), while the honey bee mite *Varroa* must develop on young (Kraus *et al.* 1986). Transmission to such specific classes of host individuals depends on colony organization: in particular, on age-dependent division of labour. For example, when infection of a worker occurs early in life and growth of the pathogen within the host is fast, the parasite is transmitted by young workers that spend most of their time attending the tasks within the nest, such as brood care. When infection is late or parasite development is slow, the infection will be spread by older workers that spend most of their time foraging outside the nest. This timing thus affects the rate of transmission to other colonies in the population (horizontal transmission). Such transmission could occur when foragers get lost or intrude into other colonies (as is the case in nectar-robbing bees), or when transmission is linked with resource use.

The latter has recently been demonstrated in the case of bumble bees and their trypanosome parasite *Crithidia bombi*: Durrer & Schmid-Hempel (1994) experimentally showed that foragers become infected when they visit a flower that has previously been visited by an infected worker. *C. bombi* is common in natural populations of its host *Bombus* spp. (Shykoff & Schmid-Hempel 1991*b*). It causes a retardation of ovarian development (Shykoff & Schmid-Hempel 1992) and a loss in the production of daughter queens for a given size of the colony (P. Schmid-Hempel & C. B. Müller, unpublished observations). Workers, males and queens become infected by ingestion of infective cells. These can readily be picked up inside the nest through contact with contaminated larvae and nest material. Wu (1994) further showed that colonies of the bumble bee *B. terrestris* are infected by several strains of *C. bombi* at the same time. In standard medium, and presumably also in hosts (Wu 1994), these strains differ in their growth rate. Fast strains should thus easily outcompete slow strains within a single host. Schmid-Hempel & Schmid-Hempel (1993) hypothesized that strain variability might nevertheless be maintained if: (i) fast strains were generally less infective to new hosts, i.e. had a disadvantage in between-host transmission; or (ii) if workers from different colonies vary in the time when they are most infectious to others. The experimental tests showed that expectation (i) was not met, because transmission shortly after infection (1–3 d) was always more successful than transmission at a later time (4–12 d after infection). However, expectation (ii) was supported, because the time period after infection during which transmission to new hosts was most effective varied significantly among colonies, with some colonies acting as 'filters' for short transmission intervals and others for longer ones (Schmid-Hempel & Schmid-Hempel 1993). Hence, strain variability could be maintained through colony-level variation in the most effective transmission interval after infection.

(b) Mating systems

The breeding system of social insects has received increased attention because it provides information basic to the understanding of social evolution and speciation (see, for example, Ward 1989) and the evolution of sex ratios in natural populations (Boomsma & Grafen 1991). The current hypothesis is that sociality has evolved because of the effects of kin selection (Hamilton 1964). It is therefore expected that individuals within groups of socially living animals, such as workers of social insect colonies, are typically more closely related to each other than the population average. In social Hymenoptera the haplodiploid reproductive system can indeed ensure high degrees of relatedness. However, genetic evidence shows that social insect colonies often do not consist of closely related individuals as a consequence of polyandry and polygyny. The latter includes primary polygyny, foundress associations, and secondary polygyny (when additional females are admitted to the colony).

Several hypotheses for the adaptive value of multiple mating (polyandry) have been formulated. Crozier & Page (1985) propose that polyandry may be beneficial if the resulting increase in genetic variation among workers in the colony allows the expression of a better colony phenotype. This is the case when, in a fluctuating environment, increased genotypic variation expands the range of tolerable conditions. The studies of Robinson & Page (1988) indicate that different worker patrilines within a colony differ in their response to task-specific stimuli. Polyandry thus provides the possibility to acquire the necessary alleles so that the many tasks are fulfilled that are essential for successful development and reproduction. Polyandry may minimize the genetic load associated with the production of less fertile diploid males (Page & Metcalf 1982). Furthermore, polyandry may evolve as a consequence of sperm limitation, i.e. when queens require sperm from more than one male to be capable of producing large colonies. In ants, polyandry is indeed positively correlated with colony size (Cole 1983). However, this hypothesis raises the intriguing question of why males are not capable of producing or of transferring more sperm in the first place, and why they sometimes, in polyandrous aggregates, transfer far more sperm than the queen can accommodate, as for example in honeybees (Koeniger 1991). Yet other hypotheses suggest that polyandry reduces the queen–worker conflict over sex allocation to the advantage of the queen (Queller 1993).

Polygyny has similarly been explained by the production of advantageous genotypic variation at the colony level (Crozier & Consul 1976). Other hypotheses concentrate on the evolution of polygyny as a result of high dispersal risks and/or a low probability of independent colony founding for the dispersing queens (Pamilo 1991). Polygyny may also evolve when queens are comparatively short-lived with respect to the colony such that other related queens can take over (Nonacs 1988).

Quite generally, parasites are suspected to play a role in the maintenance of genetic variability of their host populations, because the production of diverse offspring helps the host to escape the constantly coevolving parasites (Hamilton *et al.* 1990; Ladle 1992). Offspring variability takes on a special meaning when dispersal is limited and sons or daughters stay at home to form social groups. Examples include the social insects, but also many other socially living organisms, such as social mammals or colonial Bryozoa. Consequently, several authors have suggested that within-colony genotypic variability in social insects, and hence polyandry and polygyny, may be the result of selection by parasites and pathogens (Tooby 1982; Hamilton 1987; Sherman *et al.* 1988).

(c) Genotypic variability

Variation in host resistance and/or parasite virulence has a genotypic component in almost any instance so far analysed (see, for example, Wakelin & Blackwell 1988). Although less intensively investigated than other organisms, social insects are no different in this respect. In the honey bee, for example, selection for increased resistance is possible against the microsporidian *Nosema* (Rinderer *et al.* 1983), acarine disease, hairless-black syndrome, or American foulbrood (Tanada & Kaya 1993). In addition, honey bees show natural variation in resistance, for example against the mite *Acarapis woodi*, foulbrood and microsporidia (Bailey & Ball 1991). For the corresponding pathogens and parasites, on the other hand, little is known about genotypic variation in infectivity or virulence.

Genotypic variation for resistance–virulence traits and genotype–genotype interactions are necessary preconditions for the parasite hypothesis of polyandry or polygyny to hold true. To test whether multiple mating actually reduces the effects of parasitic infection is straightforward in principle. However, to date only Shykoff & Schmid-Hempel (1991*c*) have tested whether the genotypic mix of bumble bee worker groups affects the spread of the parasite *Crithidia bombi*. The results demonstrated that transmission among relatives was indeed more likely than among non-relatives (figure 1*a*), suggesting that heterogeneous groups are less at risk. In addition, workers from nests that were naturally infected were more likely to become infected than those coming from nests that were not naturally infected, suggesting natural variation in susceptibility (figure 1*b*).

Further experiments support the idea that increased transmission among relatives is due not to behavioural interactions or any other kin effect, but to the matching of parasite strains and host genotypes involved. Shykoff (1991) compared transmission of *C. bombi* to relatives and non-relatives of bumble bee workers from two different nests infected with the same parasite. There was no difference in transmission success to related compared with unrelated target workers when confronted with the same strain of parasite. However, transmission success varied with the nest of origin of the parasite strain, indicating differences in the type of infection among colonies. More recently, Wu (1994) in a preliminary study found a colony–strain interaction effect in the survival of infected workers. Hence, in bumble bees infected by *C. bombi*, it is possible that a variety of parasite strains trickle into each colony through the foraging activities of the workers, for example with nectar brought

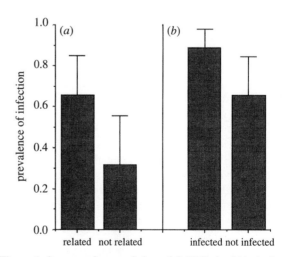

Figure 1. Success of transmission of *Crithidia bombi* in its host *Bombus terrestris* in relation to whether (*a*) the source of infection is a related or a non-related individual; (*b*) the exposed individual was from a colony that had or had not naturally acquired an infection. Bars denote average prevalence (95% c.i.) in worker groups exposed to the infection; small figures are numbers of groups tested. The differences are significant at $p < 0.05$ (from Shykoff & Schmid-Hempel 1991*c*).

back by the foragers (Durrer & Schmid-Hempel 1994). The particular genotypic mix of each colony will allow only certain strains to establish successfully; this would render polyandry potentially adaptive. In addition, the process must lead to an assortment of pathogen strains among colonies and creates the relatedness effect for the success of transmission (Shykoff & Schmid-Hempel 1991a).

3. THEORETICAL CONSIDERATIONS

(a) Polyandry

Consider a situation with ubiquitous horizontal transmission such that k different pathogen strains can infect colonies of hosts, and where the hosts have k different, corresponding resistance alleles. Colonies are supposed to contain a single queen mated randomly with a total of m males. Such a mating system will produce a mixture of homozygous and heterozygous workers in the colony. If the colony is infected by the corresponding pathogen strain (i.e. virulent with respect to the present host alleles), homozygotes are assumed to suffer and contribute less to the growth, survival and eventual reproduction of the colony with factor $w = 1 - s(w < 1)$, where s is the selection exerted by the pathogens against their individual hosts. Heterozygotes and those not infected by a virulent strain are assumed to contribute with factor 1. Imagine that these contributions affect the state, X, of the colony (e.g. its size) at time of reproduction. State X relates to fitness through a monotonically increasing 'fitness function', indicating the number of sexual individuals produced and their mating chances. As shown in the Appendix, with these assumptions, the expected state of a colony at time of reproduction is

$$E(X) = 1 - \frac{sp}{2} \text{ and } Var(X) = \frac{1}{4}\frac{s^2}{m}pq, \quad (1)$$

where $p = 1 - q =$ the probability that a queen mates with a male having an allele identical to one of hers.

The fitness function is assumed to be the result of processes operating more or less independently of the parasites' effects. For example, large colonies may be disproportionately successful because they are well defended against predators, better able to survive drought periods, or ergonomically more efficient. For simple functions, fitness can be readily calculated. (i) Linear function: $W_{\text{linear}} = E(X) = 1 - sp/2$; fitness will be indifferent to mating number. (ii) Quadratic fitness function (see figure 2): $W_{\text{quadratic}} = E(X^2) = (s^2/4m)pq + (1 - sp/2)^2$. This will select against multiple mating with factor $1/m$. (iii) Truncation selection: colonies below a certain threshold size, $X < X_0$, are assumed to produce no reproductives, while those above all have fitness 1. This is the explicit case considered by Sherman *et al.* (1988). Fitness is proportional to the variance of the distribution as given by equation (1). Hence, multiple mating will be favoured if $X_0 < E(X)$, and its benefits increase with $1/m$, because this decreases variance and less is truncated. Conversely, multiple mating will be

at a disadvantage if $X_0 > E(X)$, because this reduces the tail of the distribution that is escaping truncation.

In the general case, fitness W scales with colony state X as $W = X^a$, where a is a shape parameter. Furthermore, the number of infections per colony (by the same or different strains) is now a random variable. Mating can be costly too: females in search of mates become exposed to predators and hazards of the weather, or simply have to expend a lot of time and energy. At least for the model situation, the costs of multiple mating need not only refer to actual costs incurred when mating. The potential for ergonomic inefficiency as a result of the resulting within-colony conflict, for example, would also qualify (Schmid-Hempel 1990). The appendix shows that for this general case, the expected fitness from mating strategy m is:

$$W_m = e^{-c(m-1)}\left(e^{-\mu f} + (1 - e^{-\mu f}) \right.$$
$$\left. \times \sum_{i=0}^{m} \left(1 - \frac{i}{2m}s\right)^a \binom{m}{i} p^i q^{m-i} \right) \quad (2)$$

where $\mu =$ average number of infections per colony (a Poisson variable), $c =$ reduction of fitness for each additional mating, and $f =$ frequency of a particular, dangerous strain among the parasites. This equation has been evaluated numerically for a range of values of the parameters s, c, p, k, μ, and m. Figure 3 shows the characteristic result. Multiple mating is advantageous under conditions where costs of mating are low relative to parasite pressure (as indicated by a low ratio c/μ). More interestingly, multiple mating is favoured for values of a close to 0.5. The advantage of multiple mating is thus most evident when the shape of the fitness function implies diminishing returns from size, or in other words when colonies of intermediate size have a large share of the total reproductive

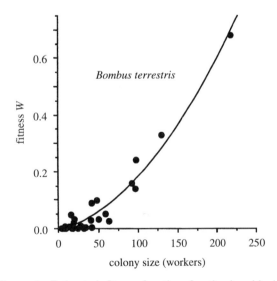

Figure 2. Estimated fitness function for the bumble bee *Bombus terrestris*, calculated according to the Shaw–Mohler equation from the number of sons and daughters produced in an experimental field population (from C. B. Müller & P. Schmid-Hempel, unpublished observations).

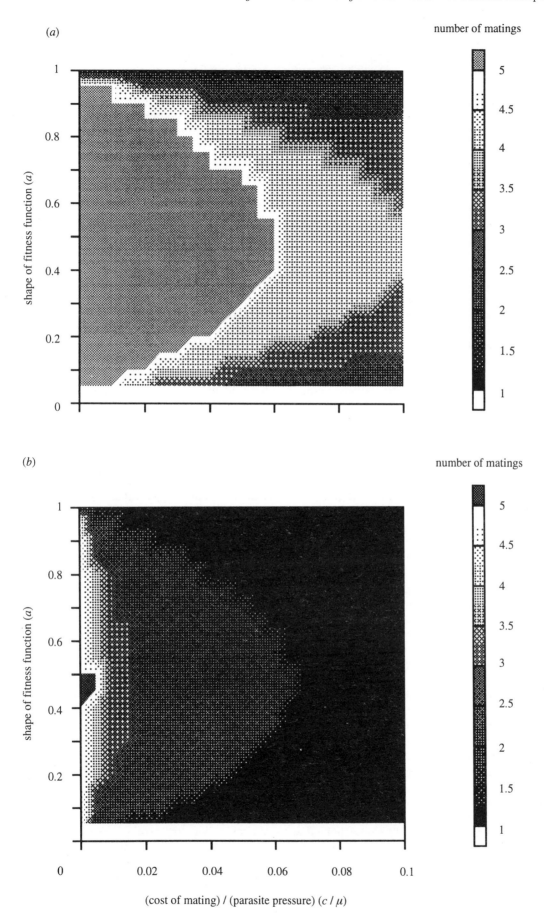

Figure 3. Number of matings (graded grey values) that yield maximum fitness as a function of the shape parameter of the fitness function (a), the costs per mating (c) and the average number of infections per colony (μ). Selection coefficient for both graphs, $s = 0.9$. (a) Small allelic diversity ($k = 4$); (b) large allelic diversity ($k = 16$).

success in the population. This might occur, for example, because of limitation in the size of available nesting sites. Note also that if multiple mating is advantageous, the best number of matings is relatively low.

In the numerical analyses, further assumptions can be relaxed. For example, what happens if queens are allowed to mate with a random number of males, m, and colonies are subject to a Poisson-distributed number of infections, of which any number of strains, $0 \ldots n$, could be virulent (i.e. when a strain matches an allele among the different worker lines)? For these cases, numerical simulations based on populations of 500 colonies were run. The results suggested the same qualitative pattern.

(b) Polygyny

Polygyny is somewhat different from polyandry because associating with other females must increase rather than decrease the risk of contracting a disease for the colony. Hence, polygyny must be advantageous for other reasons. The requirement is that each queen within a multiple female association would on average enjoy a greater success than singly nesting queens. This could occur as a result of nest site limitation or predation pressure on incipient colonies. Given that polygyny evolves, parasites could counteract the kin-selected advantages of associating with related females for the same reasons as discussed before, i.e. increased transmission as a result of genetic homogeneity of the colony. Relatedness under polygyny does indeed vary considerably. In some epiponine wasps, for example, a high degree of relatedness is maintained by cyclical oligyny (Hughes *et al.* 1993), but in *Ropalidia* serial polygyny leads to relatively low degrees of relatedness (Gadagkar *et al.* 1993), whereas the multiple queens in the ant *Solenopsis invicta* seem not to be related at all (Ross 1993).

The simplest formal treatment of the polygynous case assumes that the expected fitness per queen is an increasing function, at least up to some point, of the number of females (F) present in the colony. If queens are not related to each other and are each mated to a single, unrelated male, the probability that a given queen has mated with a male that carries an allele matching one of the queen's is again $p = 2/k$. Among F queens in random association, the expected number of queens with matching males is $E(F') = Fp$, with variance $\mathrm{Var}(F') = Fpq$, where $q = 1 - p$ as before. The expected frequency of homozygous workers in the colony is then $p/2$, that of heterozygotes $1 - (p/2)$. Similar to polyandry, equation (1) would hold if m were replaced by F, and expected fitness corrected by $1/F$. It follows that concave fitness functions would increase the benefits of polygyny whereas convex fitness functions (as in figure 2) would favour an intermediate degree of polygyny. The number of queens (F) that maximizes individual fitness can be calculated by combining equations (1) or (2) with assumptions about the increase in success with the size of the association.

4. REPERCUSSIONS

Hamilton & Zuk (1982) suggested that female mating preferences would ensure that offspring were endowed with good genes to resist parasites. The present analysis illustrates a further situation where the genetic composition of the group in terms of variability immediately affects the success of its members. The case of social insects can thus stand as a model system for highly structured populations consisting of family groups.

The formal analysis suggests that the benefit of multiple mating (and polygyny) depends on the shape of the 'fitness function' that relates colony state after parasite selection (e.g. size at reproduction) to reproductive success. Although it has been noted several times that colony size correlates positively with reproductive success (Schmid-Hempel *et al.* 1993), it is the actual shape of the distribution that is effective. In particular, polyandry is selected for when fitness is a concave (decelerating) function of colony size, as independently suggested by Sherman *et al.* (1988). The explicit case considered by Sherman *et al.* (1988), i.e. truncation selection, has also in a different context been identified as favouring genetic variability in relation to parasitism (see, for example, Hamilton *et al.* 1990).

This analysis has neglected a number of aspects of standard epidemiology, such as density-dependent transmission, immunity and heterogeneities (Anderson & May 1991), for reasons of practicability and simplicity, and concentrated instead on the problem of variance reduction (Gillespie 1977). In addition, immunity in the sense of providing long-lasting protection against repeated infections is virtually absent in insects (Gupta 1986). On the other hand, mating strategies involve not only the problem of how many males to mate with, but also the choice of mates. This can lead to a non-random assortment of genes in the population; the advantages of multiple mating should be enhanced by this. Polygyny similarly involves choice of associates. So far, this has been analysed only from the perspective of kin selection and reproductive conflict (see, for example, Keller 1993). It is clear that more detailed theoretical studies will be necessary to fully understand the long-term dynamics of the system.

Some of the predictions formulated with the present analysis are amenable to empirical verification. For example, the bumble bee *B. terrestris* depicted in figure 2 seems to be a singly mating species as expected (Röseler 1973; Estoup *et al.* 1994). Such figures for the production of males and daughter queens are rarely available for the social insects. Under certain circumstances, an argument could be made that cavity-nesters would be more likely to have a concave fitness function favouring multiple mating, whereas those species nesting in the open would attain convex curves (as in figure 2). In the genus *Apis*, *A. mellifera* and *A. cerana* are cavity nesters with a range of 10–12 matings. *A. florea*, with 3–4 mates, and *A. andreniformis*, with 7–8 matings, both nest on branches in the open (Koeniger 1991; Otis 1991). These differences are thus in line with the predictions. Note that a similar expectation would result if cavity nesting were associated with increased chances to contract durable spores of diseases, because cavities might be reused by colonies more often than the corresponding open locations on branches. As durability is also expected to lead to increased virulence of the pathogen (Ewald 1993), this too would make multiple mating more advantageous for cavity nesters. In addition, geographical variation in mating frequencies might occur in regions of the parameter space (figure 3) with steep clines such that small variation in parasite pressure or mating costs could tip the balance. Hence, the parasite hypothesis would also suggest that the degree of relatedness among members of a colony should vary locally in response to parasite pressure. It remains to be verified how important parasites are relative to the other hypotheses mentioned in the introduction.

Finally, parasites and genetic diversity within colonies are intriguing issues with respect to recognition and discrimination for or against certain classes of individuals. For example, workers infected by bee paralysis elicit aggressive responses in their nestmates (Drum & Rothenbuhler 1985). Presumably, this overrides kinship. Hence, kin can be a bane or boon with respect to disease, and discrimination for or against kin is a dilemma. Diseased relatives should be evicted from colonies, but at the same time the risk of contracting novel infections by admitting aliens should be minimized. If the main purpose of discrimination is to avoid new infections, we should expect that discrimination against workers from other colonies that represent untried and potentially dangerous associations would always be stronger than within-colony discrimination against members of different patri- (under polyandry) and matrilines (polygyny), regardless of the actual degrees of relatedness. This pattern seems to emerge from many independent studies (see, for example, Carlin *et al.* 1993). It has been pointed out that the genetic polymorphism underlying kin recognition seems to be inherently unstable and may only be maintained if linked, for example, with loci under selection by parasites (Crozier 1987).

Jacob Koella, Stella Koulianos, Regula Schmid-Hempel and Jacqui Shykoff provided helpful comments. I am especially grateful to Bill Hamilton for his comments, insights and encouragement. Financial support was provided by the Swiss National Science Foundation (grant no. 31-32193.91).

REFERENCES

Alexander, R.D. 1974 The evolution of social behaviour. *A. Rev. Ecol. Syst.* **5**, 325–383.

Anderson, R.M. & May, R.M. 1991 *Infectious diseases of humans.* Oxford University Press.

Bailey, L. & Ball, B.V. 1991 *Honey bee pathology*, 2nd edn. London: Academic Press.

Bailey, L., Ball, B.V. & Perry, J.N. 1983 Honeybee paralysis: its natural spread and its diminished incidence in England and Wales. *J. apic. Res.* **22**, 191–195.

Bailey, L. & Gibbs, A.J. 1964 Acute infection of bees with paralysis virus. *J. Insect Pathol.* **6**, 395–407.

Becker, N. & Angulo, J. 1981 On estimating contagiousness of a disease transmitted from person to person. *Math. Biosci.* **54**, 137–154.

Boomsma, J.J. & Grafen, A. 1991 Colony-level sex ratio selection in the eusocial Hymenoptera. *J. evol. Biol.* **3**, 383–407.

Brown, C.R. & Brown, M.B. 1986 Ectoparasitism as a cost of coloniality in cliff swallows. *Ecology* **67**, 1206–1218.

Carlin, N.F., Reeve, H.K. & Cover, S.P. 1993 Kin discrimination and division of labour among matrilines in the polygynous carpenter ant, *Camponotus planatus*. In *Queen number and sociality in insects* (ed. L. Keller), pp. 362–401. Oxford University Press.

Cole, B.J. 1983 Multiple mating and the evolution of social behavior in the Hymenoptera. *Behav. Ecol. Sociobiol.* **12**, 191–291.

Crozier, R.H. 1987 Genetic aspects of kin recognition: concepts, models, and synthesis. In *Kin recognition in animals* (ed. J. C. Fletcher & C. D. Michener), pp. 55–74. Chichester: John Wiley & Sons.

Crozier, R.H. & Consul, P.C. 1976 Conditions for genetic polymorphism in social hymenoptera under selection at the colony level. *Theor. Popul. Biol.* **10**, 1–9.

Crozier, R.H. & Page, R.E. 1985 On being the right size: male contributions and multiple mating in social hymenoptera. *Behav. Ecol. Sociobiol.* **18**, 105–116.

Davies, C.R., Ayres, J.M., Dye, C. & Deane, L.M. 1991 Malaria infection rate of Amazonian primates increases with body weight and group size. *Funct. Ecol.* **5**, 655–662.

Drum, N.H. & Rothenbuhler, W.C. 1985 Differences in non-singing aggressive responses of worker honeybees to diseased and healthy bees in May and July. *J. apic. Res.* **24**, 184–187.

Durrer, S. & Schmid-Hempel, P. 1994 Shared use of flowers leads to horizontal pathogen transmission. *Proc. R. Soc. Lond.* B **258** (In the press.)

Estoup, A., Scholl, A., Pouvreau, A. & Solignac, M. 1994 Monandry and polyandry in bumble bees (Hymenoptera, Bombinae) as evidenced by hypervariable microsatellites. *Molec. Ecol.* (In the press.)

Ewald, P.W. 1993 The evolution of virulence. *Scient. Am.* **268**, 56–62.

Feener, D.H.J. 1987 Size-selective oviposition in *Pseudaceton crawfordi* (Diptera: Phoridae), a parasite of fire ants. *Ann. ent. Soc. Am.* **80**, 148–151.

Feener, D.H.J. 1988 Effects of parasites on foraging and defense behavior of termitophagous ants, *Pheidole titanis* Wheeler (Hymenoptera: Formicidae). *Behav. Ecol. Sociobiol.* **22**, 421–427.

Freeland, W.J. 1976 Pathogens and the evolution of primate sociality. *Biotropica* **8**, 12–24.

Gadagkar, R., Chandrashekara, K., Chandran, S. & Bhagavan, S. 1993 Serial polygyny in the primitively eusocial wasp *Ropalidia marginata*: implications for the study of sociality. In *Queen number and sociality in insects* (ed. L. Keller), pp. 188–214. Oxford University Press.

Gillespie, J.H. 1977 Natural selection for variances in offspring number: A new evolutionary principle. *Am. Nat.* **111**, 1010–1014.

Gupta, A.P. (ed.) 1986 *Hemocytic and humoral immunity in arthropods.* New York: John Wiley & Sons.

Hamilton, W.D. 1964 The genetical evolution of social behavior. *J. theor. Biol.* **7**, 1–52.

Hamilton, W.D. 1987 Kinship, recognition, disease, and intelligence: constraints of social evolution. In *Animal societies: theory and facts* (ed. Y. Ito, J. L. Brown & J. Kikkawa), pp. 81–102. Tokyo: Japanese Scientific Society Press.

Hamilton, W.D., Axelrod, A. & Tanese, R. 1990 Sexual reproduction as an adaptation to resist parasites (a review). *Proc. natn. Acad. Sci. U.S.A.* **87**, 3566–3573.

Hamilton, W.D. & Zuk, M. 1982 Heritable true fitness and bright birds: a role for parasites? *Science, Wash.* **218**, 384–387.

Hughes, C.R., Queller, D.C., Strassmann, J.E., Solis, C.R., Negron-Sotomayor, J.A. & Gastreich, K.R. 1993 The maintenance of high genetic relatedness in multi-queen colonies of social wasps. In *Queen number and sociality in insects* (ed. L. Keller), pp. 151–170. Oxford University Press.

Jouvenaz, D.P., Lofgren, C.S. & Allen, G.E. 1981 Transmission and infectivity of spores of *Burenella dimorpha* (Microsporidae, Burenellidae). *J. invert. Pathol.* **37**, 265–268.

Keller, L. (ed.) 1993 *Queen number and sociality in insects.* Oxford University Press.

Koeniger, G. 1991 Diversity in Apis mating systems. In *Diversity in the genus Apis* (ed. D. R. Smith), pp. 199–212. Boulder, Colorado: Westview Press.

Kramm, K.R., West, D.F. & Rockenbach, P.G. 1982 Termite pathogens: transfer of the entomophagous *Metarhizium anisoplia* between *Reticulitermes* sp. termites. *J. Invert. Pathol.* **40**, 1–6.

Kraus, B., Koeniger, N. & Fuchs, S. 1986 Unterscheidung zwischen Bienen verschiedenen Alters durch *Varroa jacobsoni* Oud. und Bevorzugung von Ammenbienen im Sommervolk. *Apidologie* **17**, 257–266.

Ladle, R.J. 1992 Parasites and sex: catching the red queen. *Trends Ecol. Evol.* **7**, 405–408.

Low, B.S. 1988 In *Human reproductive behaviour: a Darwinian perspective* (ed. L. Belzig & M. Bergerhoff-Mulder), pp. 115–127. Cambridge University Press.

Nonacs, P. 1988 Queen number in colonies of social hymenoptera. *Evolution* **42**, 566–580.

Otis, G. 1991 A review of the diversity of species within Apis. In *Diversity in the genus Apis* (ed. D. R. Smith), pp. 29–49. Boulder, Colorado: Westview Press.

Page, R.E.J. & Metcalf, R.A. 1982 Multiple mating, sperm utilization, and social evolution. *Am. Nat.* **119**, 263–281.

Pamilo, P. 1991 Evolution of colony characteristics in social insects. 2. Number of reproductive individuals. *Am. Nat.* **137**, 83–107.

Poulin, R. & Fitzgerald, G.J. 1989 Shoaling as an anti-ectoparasite mechanism in sticklebacks. *Behav. Ecol. Sociobiol.* **24**, 251–255.

Queller, D.C. 1993 Worker control of sex ratios and selection for extreme multiple mating by queens. *Am. Nat.* **142**, 346–351.

Rinderer, T.E., Collins, A.M. & Brown, M.A. 1983 Heritabilities and correlations of the honey bee: response to *Nosema apis*, longevity, and alarm response to isopentyl acetate. *Apidologie* **14**, 79–85.

Robinson, G.E. & Page, R.E.J. 1988 Genetic determination of guarding and undertaking in honey-bee colonies. *Nature, Lond.* **333**, 356–358.

Röseler, P.-F. 1973 Die Anzahl Spermien im Receptaculum seminis von Hummelköniginnen (Hymenoptera, Apidae, Bombinae). *Apidologie* **4**, 267–274.

Ross, K.G. 1993 The breeding system of the fire ant *Solenopsis invicta*: effects on colony genetic structure. *Am. Nat.* **141**, 554–576.

Schmid-Hempel, P. 1990 Reproductive competition and the evolution of work load in social insects. *Am. Nat.* **135**, 501–526.

Schmid-Hempel, P. & Schmid-Hempel, R. 1993 Transmission of a pathogen in *Bombus terrestris*, with a note on division of labour in social insects. *Behav. Ecol. Sociobiol.* **33**, 319–327.

Schmid-Hempel, P., Winston, M.L. & Ydenberg, R.C. 1993 Invitation paper (The C. P. Alexander Fund): The foraging behavior of individual workers in relation to colony state in the social hymenoptera. *Can. Ent.* **126**, 129–160.

Sherman, P.W., Seeley, T.D. & Reeve, H.K. 1988 Parasites, pathogens, and polyandry in social Hymenoptera. *Am. Nat.* **131**, 602–610.

Shykoff, J.A. 1991 *On genetic diversity and parasite transmission in socially structured populations.* Ph.D. Thesis, University of Basle.

Shykoff, J.A. & Schmid-Hempel, P. 1991a Genetic relatedness and eusociality: parasite-mediated selection on the genetic composition of groups. *Behav. Ecol. Sociobiol.* **28**, 371–376.

Shykoff, J.A. & Schmid-Hempel, P. 1991b Incidence and effects of four parasites in populations of bumble bees in Switzerland. *Apidologie* **22**, 117–125.

Shykoff, J.A. & Schmid-Hempel, P. 1991c Parasites and the advantage of genetic variability within social insect colonies. *Proc. R. Soc. Lond.* B **243**, 55–58.

Shykoff, J.A. & Schmid-Hempel, P. 1992 Parasites delay worker reproduction in the bumblebees: consequences for eusociality. *Behav. Ecol.* **2**, 242–248.

Tanada, Y. & Kaya, H.K. 1993 *Insect pathology.* San Diego, California: Academic Press.

Tooby, J. 1982 Pathogens, polymorphism, and the evolution of sex. *J. theor. Biol.* **97**, 557–576.

Waddington, K.D. & Rothenbuhler, W.C. 1976 Behaviour associated with hairless-black syndrome of adult honeybees. *J. apic. Res.* **15**, 35–41.

Wakelin, D. & Blackwell, J.M. (eds) 1988 *Genetics and resistance to bacterial and parasitic infections.* London: Taylor & Francis.

Wang, D.I. & Moeller, F.E. 1970 The division of labor and queen attendance behavior of *Nosema*-infected worker honey bees. *J. econ. Entomol.* **63**, 1539–1541.

Ward, P.S. 1989 Genetic and social changes associated with ant speciation. In *The genetics of social evolution* (ed. M. D. Breed & R. E. Page), pp. 123–148. Boulder, Colorado: Westview Press.

Wilson, E.O. 1971 *The insect societies.* Cambridge, Massachusetts: Harvard University Press.

Wu, W. 1994 *Microevolutionary studies on a host-parasite interaction.* Ph.D. Thesis, University of Basle.

APPENDIX

The worker's contribution is to increase the size (or state) of the colony, X, each proportional to the effort she is able to give. However, for convenience the colony sizes reached if all workers are uninfected or infected are scaled to unity and to $w = 1 - s$ respectively, s being a measure of the average fractional weakening caused by the pathogens to a susceptible host. Each colony is exposed to a random sample of the k pathogen strains, but at most one of the infections is assumed to be dangerous for the homozygotes (this latter assumption can be relaxed in numerical simulations).

The weakening caused to individual workers is presumed to be additive in reducing the size of the colony. Each queen through her polyandrous mating can be considered to be sampling the genes of the population at random. Under the assumed additivity the mean achievement of queens building their colonies, $E(X)$, will obviously be independent of the number of males (m) they sample and will depend on (i) the chance that a sampled male has an allele the queen carries (p) (the number of alleles, k, in the population being large, the frequency of homozygous queens are taken as negligible), and (ii) the independent chance that the queen's gamete contributed to the worker carries the same allele as the male ($1/2$). Mean colony size is therefore $E(X) = 1 - sp/2$.

The variance of achievements, $\mathrm{Var}(X)$, however, reduces with m. If colony sizes are large and queens mate with single males, the variance of colony size is that of a binary variate taking values 1 and $1 - s/2$ with frequencies $q = 1 - p$ and p; thus $\mathrm{Var}(X) = (s^2/4)pq$. For the case where the queen mates with m males, the variance of the mean of m such varying sets is therefore $\mathrm{Var}(X) = (s^2/4m)pq$. For the case of quadratic dependence of colony fitness on colony size we need: $E(X^2) = \sum_i X_i^2 P(X = X_i)$. This expression can be found by reversing the 'correction' that provides a variance from a simple sum of squares: $E(X^2) = \mathrm{Var}(X) + [E(X)]^2$. Hence, $E(X^2) = (s^2/4m)pq + (1 - sp/2)^2$.

In the general case, when colony fitness depends on colony size with a power function, exponent a, it is easier to consider the explicit formulation. The queen's achievement when she mates with m males of which i have an identical allele to one of hers is

$$X_i = \frac{1}{2}\left(\frac{i}{m}w\right) + \frac{1}{2}\left(\frac{i}{m}1\right) + \left(1 - \frac{i}{m}1\right) = 1 - \frac{i}{2m}s.$$

Under these conditions, the expected colony fitness, W_m, is

$$W_m = E(X) = \sum_{i=0}^{m} X_i P(X = X_i)$$

$$= \sum_{i=0}^{m}\left(1 - \frac{i}{2m}s\right)^a \binom{m}{i} p^i q^{m-i}. \qquad (A1)$$

Colonies are infected by an average of independently acquired, Poisson distributed parasite strains. The probability of the colony not being infected is therefore $P_0 = e^{-\mu}$. Suppose that among the strains the frequency of a particular one (or a set of several) dangerous for a colony is f. The number of infections by this strain (or set) will also be Poisson distributed

with mean μf. Hence, the probability of a colony receiving no infections by this strain (or set) is $P_A = 1 - e^{-\mu f}$. Since infection is assumed to build up rapidly within a colony, all other cases than zero dangerous infections can be considered equivalent, i.e. the colony becomes totally infected. Moreover, $P_1 =$ probability that the colony is infected but the corresponding strain to which workers are susceptible is absent, hence $P_1 = e^{-\mu f} - e^{-\mu}$. The expected fitness for the colony is therefore: $W = (P_0 \cdot 1) + (P_1 \cdot 1) + (P_A \cdot W_m) = (e^{-\mu f}) + (1 - e^{-\mu f})W_m$. The qualitative conclusions are the same if infections are, for example, negatively binomially distributed.

Suppose, in addition, that mating has an independent cost such that a single mating is scaled to unity and each additional mating reduces fitness by a constant amount c. In this case, the expected colony fitness for a queen accepting m matings is:

$$W = e^{-c(m-1)}\left[e^{-\mu f} + (1 - e^{-\mu f})\right.$$

$$\left. \times \sum_{i=0}^{m}\left(1 - \frac{i}{2m}s\right)^a \binom{m}{i} p^i q^{m-i}\right]. \qquad (A2)$$

Dominant-recessive genetic system

In seeking the mean colony size for the whole population, recall that, in whatever colony they are produced, all worker contributions in the population are additive. Thus since susceptible aa workers have frequency q^2, to obtain the mean, the standard colony fitness (1) is decreased proportional to this: $E(X) = 1 - sq^2$. In a linear fitness function, this is also the mean colony fitness. In the quadratic fitness function, we need first to find the variance; for this we need to know the frequencies of the different kinds of female broods that can emerge from matings with a single male. Given Hardy–Weinberg frequencies, aa broods have frequency q^3, broods with 'A' phenotypes and aa in $1:1$ ratios have frequency $2pq^2$, and all other broods are entirely of 'A' phenotype, frequency $1 - 2pq^3 - q^3$. The sizes of those types of colony are $1 - s$, $1 - s/2$, and 1; on the basis of these, together with their frequencies, via some algebra, we find

$$E(X^2) = 1 + sq^2\left[s\left(\frac{1}{m}p + q\right) - 2\right].$$

Hence, similar conclusions would also follow for this genetic system.

6

Infectious diseases, reproductive effort and the cost of reproduction in birds

L. GUSTAFSSON, D. NORDLING, M. S. ANDERSSON, B. C. SHELDON AND
A. QVARNSTRÖM

Department of Zoology, Uppsala University, Villavägen 9, S-752 36 Uppsala, Sweden

SUMMARY

Reproductive effort can have profound effects on subsequent performance. Field experiments on the collared flycatcher (*Ficedula albicollis*) have demonstrated a number of trade-offs between life-history traits at different ages. The mechanism by which reproductive effort is mediated into future reproductive performance remains obscure. Anti-parasite adaptations such as cell-mediated immunity may probably also be costly. Hence the possibility exists of a trade-off between reproductive effort and the ability to resist parasitic infection. Serological tests on unmanipulated collared flycatchers show that pre-breeding nutritional status correlates positively with reproductive success and negatively with susceptibility to parasitism (viruses, bacteria and protozoan parasites). Both immune response and several indicators of infectious disease correlate negatively with reproductive success. Similar relations are found between secondary sexual characters and infection parameters. For brood-size-manipulated birds there was a significant interaction between experimentally increased reproductive effort and parasitic infection rate with regard to both current and future fecundity. It seems possible that the interaction between parasitic infection, nutrition and reproductive effort can be an important mechanism in the ultimate shaping of life-history variation in avian populations.

1. INTRODUCTION

That reproduction is costly constitutes a core assumption in life history theory (Williams 1966*a*,*b*). The idea that reproduction competes with other functions within the individual is as old as biology itself and can be illustrated by the idea of a Darwinian demon. A Darwinian demon can be characterized as an organism that produces an infinite number of offspring, lives for ever and as a consequence does not recognize any cost of reproduction or of living (we are among many that would envy such an organism).

A Darwinian demon simultaneously maximizes life-history traits such as longevity, breeding frequency, size at birth, rate of growth, and number and size of offspring. However, life-history theory predicts that organisms should face a trade-off between investment in current and future reproduction (Williams 1966*a*,*b*). This would be manifested via several intermediate trade-offs, such as current reproduction and survival, current and future reproduction, and number and quality of offspring.

Trade-offs can be either physiological or evolutionary. By physiological trade-offs we mean the allocation between two processes competing for the same limited resources within a single individual. Evolutionary trade-offs are genetically based and

can be demonstrated by selection experiments on phenotypes and the correlated response in subsequent generations. Physiological trade-offs as well as ecologically mediated trade-offs (see, for example, Partridge 1987) are the background of all evolutionary trade-offs; this paper mainly considers physiological trade-offs.

Orton (1929) suggested that the physiological costs of reproduction might explain the occurrence of natural death and senescence. When reproductive investment and the associated physiological costs become too high, the organism dies as a result of the deterioration of the body. Jönsson & Tuomi (1994) state that the potential for the physiological costs of reproduction having an evolutionary significance was first indicated by Fisher (1930), who pointed out two evolutionarily interesting aspects of trade-offs. Fisher was interested in the physiological mechanisms that allocate nutrition to either reproduction or maintenance and also under what circumstances one might find specific life histories.

The occurrence of trade-offs is not mysterious, but actually rather trivial. In life-history theory, the important issue is not whether any relations exist, as some are inevitable, but rather, which of the possible combinations occur and are strongest. Stearns (1992) discusses 45 possible trade-offs based on ten life-history traits. Precisely which combinations occur will depend on the interaction of selection pressures over evolutionary time and on the precise way in which physiological processes operate in a specific environment. The problem organisms face is to optimize different combinations. There is not one single solution to the problem; it will differ between populations and environments.

In most cases trade-offs can only be demonstrated convincingly by manipulating one of the involved traits. The reason for this is that individuals differ in their resources and in how these resources are allocated. This effect has been referred to as the silver spoon effect or the big house and big car syndrome (Van Noordwijk & de Jong 1986): to those that already have, more shall be given. Therefore, we need to break up the positive correlations between the life-history traits with phenotypic experiments (Gustafsson & Sutherland 1988). A number of studies have confirmed, for example, that experimentally increasing current reproductive effort results in decreased survival, future reproduction or offspring fitness (see reviews in Lessells 1991, Stearns 1992).

It is well known that nutritional status affects fecundity (Fisher 1930; Drent & Daan 1980) and that nutritional status is also important for defence against infection (Baron 1988). As early as 1871, Darwin proposed that the health and vigour of females determined both the date at which they breed and their fecundity. It is therefore important to measure both condition and disease in life-history studies. Possibly, individuals that spend more energy on current reproduction have relatively fewer resources to devote to immune defence against parasites; or individuals in lower condition that attempt to spend the same amount of resources on reproduction as those in higher condition are constrained to devote fewer resources to immune defence. These conjectures suggest a trade-off between reproductive effort and the ability to resist parasitic infections.

This paper reviews earlier and recent findings on reproductive effort, cost of reproduction and its link to infectious diseases from our long-term study of the collared flycatcher. We also give examples of, and point out the similarities between, effects of diseases and effort on both life-history characters and secondary sexual characters.

2. A CASE STUDY: LIFE HISTORY TRADE-OFFS IN THE COLLARED FLYCATCHER

(a) The species

The collared flycatcher has been studied since 1980 on the Baltic island of Gotland, where it forms an isolated population with a remarkably high site fidelity of both adults and young (Gustafsson 1985; Pärt & Gustafsson 1989) such that the lifetime reproductive success can be assessed with an accuracy rarely possible under natural conditions (Gustafsson 1989). Adult birds are caught with traps in the nest boxes, while feeding their young, and marked with unique numbered rings. Laying date, clutch size, hatching date, the number of hatched young and number of fledglings in each brood are recorded. Thirteen days after hatching, each young bird is uniquely ringed and its mass and tarsus length are measured. Adult survival and the number of offspring recruited into the breeding population are determined by catching all breeding pairs within the study area in subsequent years. Within the study area we have more than 900 nest boxes in nine woodlands which attract annually about 300–400 pairs of collared flycatchers. Eight deciduous woods are dominated by oak (*Quercus robur*) and ash (*Fraxinus excelsior*), with a dense understory of hazel (*Corylus avellana*) and hawthorn (*Crategus spp.*). The one area of coniferous forest is dominated by pine (*Pinus sylvestris*), with some birch (*Betula pubescens*).

The first male flycatchers arrive on Gotland in the first week of May and immediately take up territories. The latest arrive probably during the first week of June. The first females arrive, on average, one week after the earliest males. Nest building starts in early to mid-May and the first eggs are laid around 10–20 May so that most clutches hatch during the first half of June. Broods contain up to 8 young, which normally fledge on the fourteenth or fifteenth day after hatching. Most have left the nest by 5 July and the latest around 15 July. Only one clutch is laid, except for a few replacement clutches.

(b) Brood size experiments

In several different years (1983–5, 1988 and 1992–4) the broods of more than 1000 collared flycatchers were manipulated. Broods with the same hatching date were treated in pairs and either one or two young

Table 1. *The effect of experimentally increasing brood size on some life-history parameters for adult male and female flycatchers and offspring*

(Relationships are indicated as significantly positive (+), significantly negative (−) or non-significant (0); n.a., not applicable. Data are from Gustafsson & Sutherland (1988), Gustafsson & Pärt (1990), Schluter & Gustafsson (1993), L. Gustafsson & W. J. Sutherland, unpublished.)

	survival	fecundity	time of breeding	growth	age of first breeding	LRS
male	0	0	0	n.a.	n.a.	0
female	0	−	0	n.a.	n.a.	−
offspring	−	−	+	−	+	−

were, at two days of age, moved from one nest to the other, or for the control group two young were swapped between nests. The number of young added or removed was independent of the original number laid. Small-scale manipulations like these have the disadvantage that large sample sizes are needed to detect any effects, but the advantage that the achieved brood-size variation is similar to the natural heritable variation, which is crucial for a study to elucidate evolution.

The effects of manipulating brood size on life-history parameters are shown in table 1. The survival of both males and females was unaffected by the manipulation, but the subsequent fecundity of females was decreased by the manipulation. Furthermore, the brood size manipulation increased the number of fledged young. However, the growth of the nestlings was affected, such that larger broods resulted in lighter and smaller fledglings that were less likely to survive (Lindén *et al.* 1992). The combination of these trade-offs meant that females laying natural clutch sizes had the highest lifetime reproductive success (L. Gustafsson & W. J. Sutherland, unpublished). Consequently individuals seem, on average, to adjust their clutch to the optimal size.

Table 1 also gives examples of intergenerational trade-offs, that is the effect increased parental effort may have on the next generation. Both male and female offspring bred later in their life when they came from a brood that had been increased in size. Female offspring from such broods were less fecund which resulted in a reduced lifetime reproductive success for female offspring from larger broods (L. Gustafsson & W. J. Sutherland, unpublished).

Clutch size, as is the case with other life-history traits, is typically age-related (Gustafsson & Pärt 1990). It has been suggested that selection for high early fecundity is related to reduced later fecundity (Medawar 1952; Williams 1957, 1966*b*). This is supported by the observation that females that do not breed in their first year of life, and thereby have a lower early effort, have larger clutches later in life than those that bred in their first year (Gustafsson & Pärt 1990). This is confirmed by the effect of brood-size experiments on birds one year old. Those that had an increased brood had smaller clutches later in life than did the control birds. Consequently there is a trade-off between early and late reproduction (Gustafsson & Pärt 1990).

When the expectation of successful future reproduction at older ages is reduced by senescence, as it is in the collared flycatcher, life-history theory predicts that reproductive effort should increase with age. This is because, above certain ages, the trade-off between current and future reproduction should tip towards current and individuals are expected to make a terminal investment in reproduction (Clutton-Brock 1984). We (Pärt *et al.* 1992) investigated whether collared flycatchers tend to make a terminal investment in reproduction by comparing nestling feeding rates, energy expenditure (measured with the double-labelled water technique) and mass loss in females 5 years old compared with middle-aged females (e.g.

2–4 years old), which are at their peak performance in life (Gustafsson & Pärt 1990). Old females fed their young more frequently, consumed more energy and lost more mass. This increased investment resulted in higher mortality for old females with many young. It clearly demonstrates the effect of the allocation of resources between reproduction and survival, and indicates that there is a tendency to terminal investment.

3. MECHANISMS OF LIFE-HISTORY TRADE-OFFS

It is relatively easy to imagine how reproductive effort early in a season might affect effort later in the season (see, for example, Møller 1993, 1994), but how is an effort in one year mediated into a reproductive cost the following year? Figure 1 shows the annual cycle of the collared flycatcher and the experimental and sampling periods. It also illustrates that an effort during egg-laying or when feeding the young must have an impact on the bird during moulting, during the migration to Africa in winter, and during the spring return migration. Finally (if the bird has survived) this impact is expressed during the following breeding season. To explore possible mechanisms, we make experiments during the laying period and the nestling period. We take blood samples from the birds when they arrive and later during the breeding period. Serological tests give us measures of the birds' condition and health before breeding, during breeding and after our experiments during the feeding period.

(a) *Breeding time, phenotypic quality and reproductive success*

What determines individual variation in breeding time? Darwin (1871) proposed that the health and vigour of females determined both the date at which they breed and their fecundity. Fecundity and reproductive success decline with season in the collared flycatcher. The earliest birds seem to match their laying date to anticipate the peak period of food abundance. Later and less fecund birds have fledged

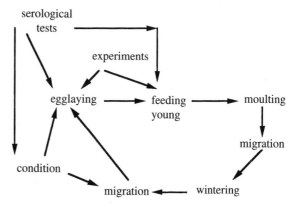

Figure 1. The annual cycle of the collared flycatcher and the experimental and sampling periods.

their young when food is less abundant (L. Gustafsson, T. Pärt & A. Lundquist, unpublished). To separate the proximate effect of time *per se* from the phenotypic quality of the parents we (L. Gustafsson *et al.*, unpublished; Wiggins *et al.* 1994) experimentally made birds hatch one week earlier or later by swapping clutches that would have hatched at different dates. If parental quality were the main factor explaining reproductive success we predicted that we should not observe a change in success compared with control birds. If the main effect were due to time *per se*, reproductive success in the experimental group would decline in parallel with control birds.

The results were intermediate. For delayed birds the reproductive success was above that of the control birds (indicating higher parental quality) but below the expected value from their breeding time (indicating a presumable environmental effect of time *per se*). The same results were obtained for the advanced group: later birds were not as successful as early birds, but their earlier hatching increased their success compared with the late controls. This demonstrates that there is both an environmental (time *per se*) and parental quality effect on the seasonal decline in reproductive success in the collared flycatcher (L. Gustafsson, T. Pärt & A. Lundquist, unpublished).

(b) Condition and reproductive success

What is this quality effect? To be able to estimate condition in the period before breeding, we measured the amount of glycosylated haemoglobin when the birds arrived at the breeding ground (Andersson 1993; M. S. Andersson & L. Gustafsson, unpublished). Hazelwood (1972) and Bairlein (1983) studied the seasonal fluctuation of blood glucose levels in migratory birds. On the basis of their results we expect birds in bad condition to stop more frequently and for longer feeding periods during migration, and to have lower average blood glucose levels. Glycosylated haemoglobin reflects the average blood glucose level during a period of two to three weeks before the blood was sampled (Andersson 1993; M. S. Andersson & L. Gustafsson, unpublished).

The proportion of glycosylated haemoglobin decreased rapidly with arrival time, indicating that late birds had been in poor condition during migration. That was further supported by a positive relation between glycosylated haemoglobin and plasma protein, which is also an indication of nutritional status (Andersson 1993; M. S. Andersson & L. Gustafsson, unpublished). Furthermore, there was a positive relation between glycosylated haemoglobin and reproductive success when breeding time was controlled for (Andersson 1993; M. S. Andersson & L. Gustafsson, unpublished). This result shows that birds in good condition not only arrived early, but also produced more offspring than expected from their breeding date. This relationship could perhaps partly explain the difference in parental quality between early and late birds.

(c) Infectious diseases and reproductive success

What causes these differences in condition? Infectious diseases have been a neglected field in evolutionary ecology (see, for example, Sheldon 1993). Much recent work on host–parasite interactions has focused on macroparasites such as haematozoa and ectoparasites, which are relatively easily quantified (Loye & Zuk 1991). The impact that parasites have on their hosts is usually expressed in terms of pathogenicity or virulence of a specific type of parasite. However, the impact of a particular parasite on its host can never be generalized, because it depends to a large extent on the quality of the host's immune system, which is highly variable. We have used an alternative way of studying parasite–host interaction simply by measuring the activity of the birds' immune systems and using this as a guide to whether any parasite is currently causing disease in each individual bird. We performed several serological tests on blood samples collected from the birds. In concert these tests function as a screening procedure that detect disease processes rather than as specific diagnostic tests (Hunter 1989; Nordling 1993).

(i) Serological tests

The total white blood cell (WBC) count is the most widely used and perhaps the single most important serological test. It offers general information about infection status and can thus function as a screening test. Many different avian species respond to a wide range of pathogenic agents with increasing WBC counts (Dein 1986). In birds, this increase has been associated with infection by macroparasites and bacteria, including tuberculosis, as well as different viruses, such as herpes. However, the total WBC count is the most difficult parameter to measure in avian blood because of the interference of lysed red blood cells (RBC) and thrombocytes. To overcome this problem, we chose an alternative way of measuring WBC in this study. The amount of WBC was estimated from the height of the buffy coat layer in a hematocrit capillary tube (Wardlaw & Levine 1983) by using digital callipers after centrifugation. A portable microscope with a halogen light source was constructed for this purpose. For some samples the proportion of different WBC types was determined from counts of blood smears. A selective increase in different types of white blood cell indicates what type of parasite is likely to be the cause of infection.

The sedimentation rate (SR) of red blood cells is not a standard measure in avian diagnostics. However, it has been experimentally verified as a useful measure of infection intensity (Sharma *et al.* 1984). SR is highly species-specific and can only be informative if red blood cells sediment through plasma at an intermediate rate. This turns out to be the case in the collared flycatcher. The speed is very similar to the human SR. The SR depends on plasma factors as well as blood-cell factors. It is enhanced in a wide range of infectious and inflammatory disease owing to increasing concentrations of one of the major acute-phase proteins (fibrinogen) and immunoglobulins.

The buffy layer (Wintrobe 1993) and the layer under it are useful areas for detection of blood parasites (G. F. Bennett, personal communication). Just above the buffy layer is the site where trypanosomes and microfilaria are encountered. In the buffy layer *Leucocytozoon* spp. and *Hepatozoon* may form a thin darker layer. The thin layer under the buffy coat may contain *Haemoproteus* and *Plasmodium* spp. (G. F. Bennett, personal communication). These parasite indicators were found to exist in the collared flycatchers in our study; blood smears from the buffy layer suggest a parasite prevalence of 60% (authors' unpublished results). The usual measurement of haematozoan parasite prevalence is made by counting parasites in ordinary blood smears. This suggests a parasite prevalence of about 20% in this population. Hence measurements based on ordinary blood smears may underestimate the prevalence of Haematozoa.

Proteins in general are not viewed as an energy store and are used only after both glycogen and lipid reserves are nearly or completely exhausted (Blem 1990). During times of heavy energy expenditure, such as feeding nestlings, the plasma protein concentration is likely to be very informative as a nutritional status parameter. Infectious and inflammatory diseases are another major cause of plasma protein degradation. Low concentrations signal disease. Standard spectrophotometric measurements were therefore made of plasma protein concentrations.

Different serum proteins involved in defence against infection, such as acute-phase proteins and immunoglobulins, were quantified by standard agarose gel electrophoresis. Typical patterns of infection determined from the plasma electrophoresis correlated strongly with our serological tests done in the field (figure 2 and authors' unpublished results).

(d) Infectious diseases and reproductive performance

That disease might negatively affect reproductive success seems obvious. More pertinent is the extent to which this mechanism is operating in natural populations. In the collared flycatcher we found that a number of measures of reproductive performance correlates significantly with all three infection parameters as well as with plasma protein concentrations in females (table 2). WBC counts, as expected, were related to parasite intensity. There was also a strong negative association between reproductive success and the immune response, including WBC counts, indicating overall that birds with low success had more infections and parasites. There was a particularly strong association for heterophil counts which indicate that bacteria may be important in this population, since bacterial infection results in increased levels of heterophils (Dein 1986).

An important determinant of reproductive success is time of breeding: first, food supply, especially caterpillar availability, declines quickly with season; second, good territories are occupied early in the season (L. Gustafsson, T. Pärt & A. Lundquist, unpublished). However, the causal basis

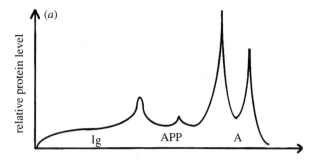

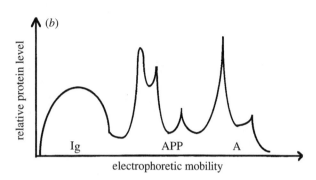

Figure 2. Schematic graph of results from standard agarose gel electrophoresis of plasma proteins performed to quantify different serum proteins (albumin A) acute-phase proteins (APP) and immunoglobulins (Ig) involved in infection defence. (*a*) A typical 'healthy collared flycatcher' (high levels of albumin, low levels of acute phase proteins and immunoglobulins): (*b*) a bird with strong signs of infectious disease (lower levels of albumin, higher levels of acute phase proteins and immunoglobulins).

for delayed breeding is poorly understood. The results in this study, with late birds showing high levels of all three infection parameters, suggest that disease might cause later breeding and also smaller clutch sizes. One can rule out the alternative suggestion that it is solely late breeding that causes disease (birds on poor territories may be exposed to more vectors, or have to look harder for food, thereby increasing susceptibility to parasites) because disease was also measured with the same result close to the arrival date, before these effects could have occurred. Because disease influences time of arrival and the start of breeding, there may also be additional physiological mecha-

Table 2. *Relations between condition and infection parameters and reproductive performance*

(Abbreviations: % HBG, proportion glycosylated haemoglobin; protein, plasma proteins; WBC, white blood cell count; SR, sedimentation rate; Ig, immunoglobulins. Data from Andersson (1993); Nordling (1993); L. Gustafsson, unpublished.)

	% HBG	protein	WBC	SR	Ig	hetero-phils
laying date	−	−	+	+	+	+
clutch size	+	+	−	0	−	−
number of fledged young	+	+	−	−	−	−

nisms, caused by disease, that directly influence clutch size. If this is the case, the nutritional constraints on clutch size and reproductive success shown to correlate with a late start to breeding (L. Gustafsson, T. Pärt & A. Lundquist, unpublished) may be caused by direct influences from disease.

(e) Diseases, reproductive effort and future reproduction

We have seen earlier that present reproductive effort can have profound effects on subsequent reproductive performance. Such reproductive trade-offs clearly must be mediated physiologically (see, for example, Calow 1979; Stearns 1992) rather than circuitously through 'ecological mechanisms'; (see, for example, Partridge 1989). The exact nature of the mechanism, however, still remains obscure (see Gustafsson 1990; Lessells 1991). Anti-parasite adaptations based on, for example, cell-mediated immunity can certainly also be costly. As a consequence, a trade-off between reproductive effort and the ability to resist parasitic infection can be expected. Some experimental evidence indicates that birds that are artificially given larger clutches are more likely to become infected (Norris *et al.* 1994). Festa-Bianchet (1989) showed that female bighorn sheep (*Ovis canadensis*) that raised male offspring or that were lactating had higher faecal counts of lungworm larvae than females that raised female offspring or that were not lactating.

Serological tests on collared flycatchers whose brood sizes were manipulated (L. Gustafsson & D. Nordling, unpublished) showed that parasitic infection rate increased with brood size. Unsurprisingly, immune response also increased, perhaps directly owing to increasing infection rate by haematozoan parasites (see, for example, Norris *et al.* 1994) but perhaps owing to other infectious diseases.

These findings suggest that the results from previous brood-size experiments in the collared flycatcher, which demonstrated a link between high reproductive effort early in life and a reduced performance later, could be explained by a link between increased effort, increased infectious disease and reduced subsequent reproduction.

A recent analysis of the effect of experimentally increased effort in one year on fecundity in the following year shows that female collared flycatchers that are parasitized by Haematozoa lay relatively smaller clutches in the year following the experiment than females that are unparasitized in the year of the experiment, where both groups receive an increase in brood size (L. Gustafsson, B. C. Sheldon, D. Nordling, F. Widemo, T. Pärt & G. F. Bennett, unpublished). Within the subset of birds that were scored as not being infected by haematozoan parasites there was no apparent effect of brood-size manipulation on future fecundity, indicating that a decline in future fecundity due to experimental manipulation arose mainly through its effect on parasitized birds. We suggest that the interaction between the immune response to

parasitism, nutritional status and the reproductive effort of birds subjected to brood-size manipulation may therefore be the mechanism through which the current versus future reproduction trade-off is mediated between years (figure 3a). This result also demonstrates that there are physiological mechanisms that can linger on throughout the annual cycle of the collared flycatcher (figure 1) and be manifested in the subsequent breeding.

4. DISEASES, REPRODUCTIVE EFFORT AND SECONDARY SEXUAL TRAITS

Parasitism and secondary sexual traits have recently been of major focus in sexual selection theory (Hamilton & Zuk 1982; Bradbury & Andersson 1987; Loye & Zuk 1991; Møller 1994). In many species, males display gaudy and elaborate ornaments, whereas females are relatively plain. It is common in these sexually dimorphic species that females show reduced expression of the same ornament as males (Darwin 1871). There is now general agreement that these display characters evolved in response to sexual selection (Bradbury & Andersson 1987; but see Alatalo *et al.* 1994). When a secondary sexual trait is costly to its bearer, it is possible that the causal path diagram of the factors affecting the expression of the trait is of a similar structure to that for life-history traits (figure 3a). For instance, diseases and condition may affect the size of the trait. The size of the trait affects effort, which affects the prevalence to diseases. This in turn affects the subsequent size of the trait (figure 3b).

(a)

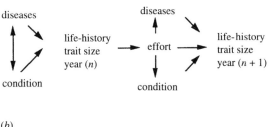

(b)

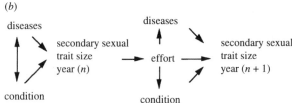

Figure 3. Causal path diagram of the effects on the expression of life-history (*a*) or secondary sexual (*b*) traits. We suggest that the interaction between infectious disease, nutritional status and the reproductive effort of birds subjected to brood-size manipulation may be the mechanism through which the current against future reproduction trade-off is mediated between years. We also suggest that the causal pathway for life-history traits is paralleled in secondary sexual traits where the same mechanisms might occur: diseases and condition affect the size of the trait and the size of the trait affects effort which in turn affects the prevalence to diseases. Finally those affect the subsequent size of the trait.

Male collared flycatchers apparently use the white areas of their plumage as signals. The white forehead patch is displayed in male–male interactions and the extensive white patches on the wings are displayed when the male tries to attract a female to the nest hole. These white patches (badges) seem to be a reliable signal of male quality: males with large badges survive better and produce more recruited offspring (table 3). Birds in good condition have large white badges; the size of the badges correlates positively with the amount of plasma protein (table 3). Birds that have large white badges also seem to be less affected by parasites; the size of the patches correlates negatively with the WBC count (table 3). In this context a Darwinian demon should be in such good condition, and so healthy, that it would have turned completely white, and be able to displace all other males and to attract all females in the world! This is, of course, not the case. What are the trade-offs that prevent this from happening?

Does a large badge induce more effort? When we enlarged and reduced the badge in a paint experiment (A. Qvarnström, unpublished) there was a significant difference in the sedimentation rate of the red blood cells later in the breeding season. The males with enlarged badges had a higher sedimentation rate, which seems to indicate that they had developed more infections than the males with reduced badges. Interestingly, in unmanipulated males one year old there was a negative relation between the proportion of glycosylated haemoglobin and badge size, indicating that males one year old with large badges are in bad condition. Furthermore, such males also displayed a positive correlation between badge size and amount of WBC; these results also indicate that they had been exposed to more infections. This was not the case for males older than one year (two to four years), for which there was no relation between badge size and percentage glycosylated haemoglobin but a negative relation between badge size and amount of WBC. These results indicate that males one year old with large badges, as well as birds with experimentally increased badges, were having to expend more effort, possibly in male–male interactions.

This was also the case for males approaching the end of their life (those more than four years old). For these males there was a significant positive relation between badge size and amount of WBC, and a negative relation between badge size and percentage glycosylated haemoglobin, indicating that these males also apparently had too large a badge for their

condition. This may be a parallel case to the terminal reproductive effort in relation to brood size discussed earlier (Pärt *et al.* 1992). In all three cases (young males with big patches, experimentally increased patches, ageing males with big patches) these relationships seem to indicate that these birds are displaying too large a badge for their condition, which possibly could result in increased morbidity and subsequent mortality. It is especially interesting that these results might offer a mechanistic explanation for how the honesty of the trait is maintained.

5. CONCLUSION AND PERSPECTIVES

The discussion of the effect of diseases on life-history traits and secondary sexual characters in an evolutionary context requires that there be some heritability of resistance to infections for natural selection to work on. Almost every organism must experience an infectious disease at least some time during its life (Price 1980). Defence against infections, including acquired immunity, can be modulated by environmental factors such as nutrition, stress and age. However, the genetic influence on the infection defence system, such as that resulting from the expression of major histocompatibility (MHC) haplotypes, sets the ultimate limits of reaction patterns in every individual. Accordingly, specific MHC alleles have been shown to be associated with resistance against disease (Van Eden *et al.* 1983; Hedrick 1994). One classic example concerns chickens, in which individuals with one specific haplotype have a much higher resistance to Marek's disease, a type of viral leukaemia (Briles *et al.* 1977). In West African children, malaria has selected for an unusually high incidence of rare MHC genes, which provide protection from severe malaria equal to the protection afforded by the sickle-cell haemoglobin variant (Hill *et al.* 1991). Some MHC B genes produce resistance to the bacterium causing fowl cholera (Lamont *et al.* 1987). The genetic variance in MHC haplotypes could be maintained by the cyclic nature of host–parasite interaction (Hamilton 1982; Anderson & May 1985; Hedrick 1994). The conclusion is that high immunological quality may offer a strong selective advantage with evolutionary consequences. Work in progress on diseases in the collared flycatcher involves the study of how MHC haplotypes interact with disease occurrence and its consequences, i.e. natural selection of genotypic variance in MHC.

As a first step towards demonstrating inheritance of

Table 3. *Associations between white plumage patterns and viability variables in male collared flycatchers*

(Abbreviations as for table 2; x, not measured separately. Data from L. Gustafsson, D. Nordling, M. Andersson & A. Qvarnström, unpublished.)

amount of white	survival	no. recruits	protein	% HBG	WBC
all males	+	+	+	0	−
1 year old	x	x	x	−	+
2–4 years old	x	x	x	0	−
5–7 years old	x	x	x	−	+

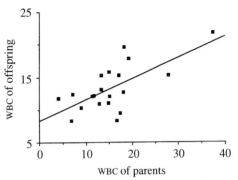

Figure 4. The parent–offspring regression of amount of white blood cells (wbc): $y = 8.30 + 0.33x$, $r^2 = 0.45$, $p < 0.001$. wbc was measured in the birds as adults.

the immune response, figure 4 shows the parent–offspring regression of amount of wbc. The amount of wbc was measured in the offspring during their first year of breeding. There is a surprisingly strong resemblance between offspring and their parents suggesting a simple heritability of *ca.* 0.66. However, beside the genetic contribution to the resemblance it must be borne in mind that there will probably also be maternal effects (Schluter & Gustafsson 1993) and possibly also common environmental effects (Gustafsson & Merilä 1994) that inflate this estimate.

In conclusion, it seems possible that the interaction between infectious diseases, nutrition and reproductive effort can be an important mechanism in the ultimate shaping of both life-history variation and secondary sexual traits in avian populations.

We thank the vast number of people that helped with field and lab work throughout this study. This study was supported by grants from the Swedish Natural Research Council to L.G.

REFERENCES

Anderson, R.M. & May, R.M. 1985 Epidemiology and genetics in the coevolution of parasites and hosts. *Proc. R. Soc. Lond.* B **219**, 281–283.

Andersson, M.S. 1993 Migratory condition affects reproductive success in the collared flycatcher, evidence from glycosylated haemoglobin. Master's thesis, University of Uppsala.

Bairlein, F. 1983 Seasonal variations of serum glucose levels in a migratory songbird, Sylvia borin. *Comp. Biochem. Physiol.* **76**, 397–399.

Baron, R.B. 1988 Protein-calorie undernutrition. In *Textbook of medicine* (ed. J. B. Wyngaarden & L. Smith), pp. 1212–1215. London: W. B. Saunders Co.

Bradbury, J.W. & Andersson, M. 1987 *Sexual selection: testing the alternatives.* New York: Wiley.

Briles, W.E., Stone, H.A. & Cole, R.K. 1977 Marek's disease: effects of B histocompatibility alloalleles in resistant and susceptible chicken lines. *Science, Wash.* **195**, 193–195.

Calow, P. 1979 The cost of reproduction – a physiological approach. *Biol. Rev.* **54**, 23–40.

Clutton-Brock, T.H. 1984 Reproductive effort and terminal investment in iteroparous animals. *Am. Nat.* **123**, 212–229.

Darwin, C. 1871 *The descent of man and selection in relation to sex.* London: Murray.

Dein, J. 1986 Hematology. In *Clinical avian medicine and surgery* (ed. G. J. Harrison & W. R. Harrison), pp. 174–191. London: Saunders Co.

Drent, R.H. & Daan, S. 1980 The prudent parent; energetic adjustments in avian breeding. *Ardea* **68**, 225–265.

Festa-Bianchet, M. 1989 Individual differences, parasites and the cost of reproduction for bighorn ewes, *Ovis canadensis. J. Anim. Ecol.* **58**, 785–795.

Fisher, R.A. 1930 The genetical theory of natural selection. Dover.

Gustafsson, L. 1985 Fitness factors in the Collared Flycatcher *Ficedula albicollis*. Temm. Thesis, Uppsala University.

Gustafsson, L. 1989 Lifetime reproductive success in the Collared Flycatcher. In *Lifetime reproductive success in birds* (ed. I. Newton), pp. 75–89. New York: Academic Press.

Gustafsson, L. 1990 Life-history trade-offs and optimal clutch size in relation to age in the collared flycatcher. In *Population biology in passerine birds: an integrated approach* (ed. J. Blondel, A. Goosler, J. D. Lebrenton & R. McCleary), pp. 235–246. Berlin: Springer-Verlag.

Gustafsson, L. & Merilä, J. 1994 Foster parent experiments reveals no genotype-environment affect in the external morphology of the collared flycatcher. *Heredity* **73**, 124–129.

Gustafsson, L. & Pärt, T. 1990 Acceleration of senescence in the collared flycatcher by reproductive costs. *Nature, Lond.* **347**, 279–281.

Gustasfsson, L. & Sutherland, W.J. 1988 The cost of reproduction in the collared flycatcher. *Nature, Lond.* **335**, 813–817.

Hamilton, W.D. 1982 Pathogens as causes of genetic diversity in their host population. In *Population biology of infectious diseases* (ed. R. M. Anderson & R. M. May), pp. 269–296. Berlin: Springer-Verlag.

Hamilton, W.D. & Zuk, M. 1982 Heritable true fitness and bright birds: a role for parasites. *Science, Wash.* **218**, 384–387.

Hazelwood, R.L. 1972 The intermediary metabolism in birds. In *Avian biology*, vol. 2, (ed. D. S. Farner & J. R. King), pp. 471–519. London: Academic Press.

Hedrick, P.W. 1994 Evolutionary genetics of the major histocompatibility complex. *Am. Nat.* **143**, 945–964.

Hill, A.V.S., Allsop, C.E.M., Kwiatkowski, D., Anstey, N.M., Twumasi, P., Rowe, P.A., Bennett, S., Brewster, D., McMichael, A.J. & Greenwood, B.M. 1991 Common West African HLA antigens are associated with protection from severe malaria. *Nature, Lond.* **352**, 595–600.

Hunter, B. 1989 *Detection of pathogens: monitoring and screening programmes.* ICBP Technical Publication 10.

Jönnson, K.I. & Tuomi, J. 1994 Cost of reproduction in a historical perspective. *TREE* **9**, 304–307.

Lamont, S.J., Bolin, C. & Cheville, N. 1987 Genetic reistance to fowl cholera is linked to MHC-complex. *Immunogenetics* **25**, 284–289.

Lessells, C.M. 1991 The evolution of life histories. In *Behavioural ecology: an evolutionary approach* (ed. J. R. Krebs & N. B. Davies), pp. 32–68. Oxford: Blackwell.

Lindén, M., Gustafsson, L. & Pärt, T. 1992 Selection on fledgling mass in the collared flycatcher and the great tit. *Ecology* **73**, 336–343.

Loye, J.E. & Zuk, M. (eds) 1991 *Bird–parasite interactions.* Oxford University Press.

Medawar, P.B. 1952 *An unsolved problem of biology.* London: H. K. Lewis.

Møller, A.P. 1993 Ectoparasites increase the cost of reproduction in their hosts. *J. Anim. Ecol.* **62**, 309–322.

Møller, A.P. 1994 *Sexual selection in the barn swallow*. Oxford University Press.

Nordling, D. 1993 Reproductive success and health status in the collared flycatcher. Master's thesis, Uppsala University.

Norris, K., Anwar, M. & Read, A.F. 1994 Reproductive effort influences the prevalence of haematozoan parasites in great tits. *J. Anim. Ecol.* **63**, 601–610.

Orton, J.H. 1929 Reproduction and death in invertebrates and fishes. *Nature, Lond.* **123**, 14–15.

Price, P.W. 1980 *Evolutionary biology of parasites*. Princeton, New Jersey: Princeton University Press.

Pärt, T. & Gustafsson, L. 1989 Causes and consequences of breeding dispersal in the Collared Flycatcher *Ficedula albicollis*. *J. Anim. Ecol.* **58**, 305–320.

Pärt, T., Gustafsson, L. & Moreno, J. 1992 'Terminal investment' and a sexual conflict in the collared flycatcher. *Am. Nat.* **140**, 868–882.

Partridge, L. 1987 Is accelerated senescence a cost of reproduction. *Funct. Ecol.* **1**, 317–320.

Quesenberry, K. & Moroff, S. 1991 Plasma electrophoresis in Psittacine birds. *Proc. Ass. Avian Vet.* **XX**, 112–119.

Sharma, S.C., Bhatia, B.B. & Pachauri, S.P. 1984 Blood cellular and biochemical studies in chicken experimentally infected with *Toxoxara canis*. *Ind. J. Parasitol.* **8**, 159–163.

Schluter, D. & Gustafsson, L. 1993 Maternal inheritance of condition and clutch size in the collared flycatcher. *Evolution* **47**, 658–667.

Sheldon, B. 1993 Sexually transmitted diseases in birds: occurrence and evolutionary significance. *Phil. Trans. R. Soc. Lond.* B **339**, 491–497.

Stearns, S.C. 1992 *The evolution of life histories*. Oxford University Press.

Van Eden, W., Devries, R.R.P. & Van Rood, J.J. 1983 The genetic approach to infectious disease with special emphasis on the MHC. *Dis. Markers* **1**, 221–242.

Van Noordwijk, A.J. & de Jong, G. 1986 Acquisition and allocation of resources: their influence in variation in life history traits. *Am. Nat.* **128**, 137–142.

Wardlaw, S.C. & Levine, R.A. 1983 Quantitative buffy coat analysis, a new laboratory tool functioning as a screening complete blood cell count. *J. Am. med. Ass.* **249**, 617–620.

Williams, G.C. 1957 Pleiotropy, natural selection and the evolution of senescence. *Evolution* **11**, 398–411.

Williams, G.C. 1966a Natural selection, the cost of reproduction and a refinement of Lack's principle. *Am. Nat.* **100**, 687–690.

Williams, G.C. 1966b *Adaptation and natural selection*. Princeton University Press.

Wintrobe, M.M. 1933 Macroscopic examination of the blood. *Am. J. med. Sci.* **185**, 58–71.

Wiggins, D., Gustafsson, L. & Pärt, T. 1994 Seasonal decline in avian reproductive success: experiments on timing and clutch size in the collared flycatcher. *Oikos*. (In the press.)

7

Evolution of mutation rate and virulence among human retroviruses

PAUL W. EWALD

Department of Biology, Amherst College, Amherst, Massachusetts 01002-5000, U.S.A.

SUMMARY

High mutation rates are generally considered to be detrimental to the fitness of multicellular organisms because mutations untune finely tuned biological machinery. However, high mutation rates may be favoured by a need to evade an immune system that has been strongly stimulated to recognize those variants that reproduced earlier during the infection. HIV infections conform to this situation because they are characterized by large numbers of viruses that are continually breaking latency and large numbers that are actively replicating throughout a long period of infection. To be transmitted, HIVs are thus generally exposed to an immune system that has been activated to destroy them in response to prior viral replication in the individual. Increases in sexual contact should contribute to this predicament by favouring evolution toward relatively high rates of replication early during infection. Because rapid replication and high mutation rate probably contribute to rapid progression of infections to AIDS, the interplay of sexual activity, replication rate, and mutation rate helps explain why HIV-1 has only recently caused a lethal pandemic, even though molecular data suggest that it may have been present in humans for more than a century. This interplay also offers an explanation for geographic differences in progression to cancer found among infections due to the other major group of human retroviruses, human T-cell lymphotropic viruses (HTLV). Finally, it suggests ways in which we can use natural selection as a tool to control the AIDS pandemic and prevent similar pandemics from arising in the future.

1. INTRODUCTION

The generation and maintenance of heritable variation is central to evolutionary processes. Although genetic variation is generated ultimately by mutation, reductions in mutation rate are generally considered beneficial to sexually reproducing organisms because a low mutation rate appears sufficient to generate a great amount of variability. Genetic recombinations during meiosis are considered to be more beneficial than high mutation rates as a means for generating variation among offspring. This is because recombination puts together different instructions that have passed the test of time (and may help weed out inferior instructions), whereas mutations generate a very high proportion of variants with reduced fitness. The existence of proof-reading functions among DNA polymerases is evidence of the disadvantageousness of high mutation rates in multicellular organisms.

High mutation rates may be much more valuable for viruses, particularly those that infect animals with well-developed immune systems. Virologists have proposed that high mutation rates among viruses reflect adaptive mechanisms for generating variability (Holland 1993). Because replication from RNA templates is associated with high mutation rates, the coding of viral genomes using RNA has been interpreted as one mechanism through which high mutation rates are maintained (Holland 1993). The replication of DNA viruses, such as the hepatitis *B* virus, through RNA templates supports this view – RNA intermediates apparently function to introduce mutations (Holland 1993). But mutation rates among viruses vary greatly, even among RNA viruses. Virologists typically explain this variation in a proximate sense by referring to the different biochemical mechanisms through which mutations are introduced, but little attention has been directed toward ultimate reasons why different levels of mutation proneness may be favoured by natural selection.

Some of this variation in mutation rate may be attributable to genome size. As genome size increases, a high mutation rate becomes more costly to the organism because mutations are generally disadvantageous; the probability of advantageous mutations occurring in the absence of overriding disadvantageous ones would therefore be extremely small. But variation in genome size does not explain why different viruses of a given genome size have different mutation rates. To understand mutation rates, one must take into account both the genome size and the relative fitness of mutants (Eigen & Biebricher 1988).

In this paper, I focus on the fitness costs and benefits of mutations to develop a hypothesis for understanding why some viruses are more mutation-prone than others. This hypothesis represents a return to the problem of 'sex versus non-sex versus parasite' (Hamilton 1980), with a focus on the value of generating genetic variation among parasites. I then consider the variation in mutation rates among retroviruses in light of this hypothesis, suggest some tests of the hypothesis, and discuss the potential importance of this issue for controlling existing viruses and for guarding against emerging viruses.

My basic proposition is that differing selective pressures imposed by immune systems may cause the evolution of different mutation rates. High mutation rates can increase the fitness of a virus by making the mutant viruses 'look' different to the immune system. A virus with a different structure may escape destruction by an immune system that has 'learned' to recognize the parental viruses.

What aspects of a virus's exposure to the immune system should be associated with relatively great fitness benefits from mutation? I suggest that high mutation should be strongly favoured when viruses: (i) engage in both active replication at the onset of infection and delayed replication from latently infected cells within each infected host; (ii) generate persistent infections from which viral progeny can be transmitted to new hosts; and (iii) infect cell types that make the viruses vulnerable to destruction by the immune system during their reproduction in and escape from the body.

The specific argument proposes that actively replicating viruses stimulate an immune response that elevates the fitness benefits of being different from parental viruses. Any virus that reproduces after a strong immune response has been generated against parental viruses has a low chance of success if: (i) its replication exposes it to the activated immune system; and (ii) if the virus has not altered its structure to allow it to escape the immune attack. Mutations always carry the potential for untuning finely tuned biochemical machinery, but under these circumstances the costs of this untuning should be weighed against the benefits of looking different to the immune system. The balance of these fitness costs and benefits should affect the overall mutation rate to which any particular virus will evolve.

At the other end of the mutation continuum should be viruses that are minimally exposed to the specific immune responses during their reproduction in and escape from the body. Most viruses that cause acute infections should lie in the vast middle ground between these two ends of the mutation continuum. Viruses that infect without latency and in a way that exposes them to effective immunological control should benefit from looking different when they are reproducing after the immune response is generated. However, a high mutation rate that could facilitate this variation at that time would probably place them at a competitive disadvantage early during infection by untuning their replicative machinery. These viruses should therefore lie toward but not at the mutation-prone end of the continuum. Viruses that infect and are transmitted from tissues with little

exposure to the immune system should get relatively little benefit from mutation and should therefore lie toward the mutation-averse side of the continuum.

This hypothesis presumes that the viral replication systems can evolve differences in mutation rates, and hence that heritable variation in mutation rate exists. Such variation has been documented: mutation rates of envelope genes in influenza *A* viruses, for example, vary by several-fold (Suárez *et al.* 1992).

The hypothesis presented above emphasizes the need to understand the evolution of replication rate. To understand the mutation-prone end of the continuum, we must understand why some viruses generate persistent infections associated with prolonged periods of transmission, and with exposure to strongly activated immune systems during transmission. To understand the mutation-averse end of the continuum, we must determine the factors that allow a virus to infect a host and be transmitted from a host with relatively little exposure to the immune system. My approach to the problem complements that of Bell (1993), who proposed that persistent infections might be responsible for long-term deterioration of hosts, that is, for senescence. The evolution of mutation rate is important to his proposition, because high mutation rates should tend to contribute to the eventual deterioration of the hosts.

2. HUMAN IMMUNODEFICIENCY VIRUSES

Human retroviruses include viruses that lie near either end of this mutation continuum. Human immunodeficiency virus type 1 (HIV-1) lies at the mutation-prone end. HIV infections are characterized by a combination of active replication of virions and latent infection. During the first few months of infection, the actively replicating viruses generate a high density of virus in the blood; thereafter, free virus is apparently continuously cleared from the bloodstream by an effective immune response (Fauci *et al.* 1991). Viral density again increases during the last two years of infection when the immune system disintegrates. Throughout this time, at least some viruses are actively replicating, primarily in the germinal centers of lymphoid organs where as many as one out of three target cells may be infected (Embretson *et al.* 1993; Piatak *et al.* 1993).

The dynamics of latency are not well understood, but some viruses are presumably breaking their latency fairly continuously during this time. Although genetic variation in tendencies to enter latency has not been well-studied in HIV, studies of SIV have demonstrated that transactivation is influenced by differences in promotor sequences (Anderson & Clements 1992). For almost all of these latent viruses, whether or not genetic differences influence tendencies toward latency, by the time latency is broken, the person's immune system will have been stimulated to defend against viruses that had previously begun active replication. The efficient clearance of virus from the bloodstream during the bulk of an HIV infection and the turnover of antigenic types during this time indicate that the immune system is exerting a strong selective pressure on viruses that are replicating during the years between the initial immune response and the eventual disintegration of the immune system.

A high mutation rate therefore probably increases the chances of diverging from the earlier viruses sufficiently to escape destruction by the immune system (Albert *et al.* 1990; Coffin 1990; Tremblay & Wainberg 1990). Indeed, direct measurements of mutations and reverse transcription show that HIV's reverse transcriptase tends to generate mutations at greater rates than the reverse transcriptases of most other retroviruses (Preston *et al.* 1988; Roberts *et al.* 1988; Takeuchi *et al.* 1988; Bakhanashvili & Hizi 1992*a*,*b*; Hübner *et al.* 1992; Ji & Loeb 1992; Monk *et al.* 1992; Varela–Echavarria *et al.* 1992).

The relevance to human infections of these high mutation rates has been questioned by Temin (1989*a*) because such high rates would make virtually all of the progeny viruses from an infected cell different from the parental virus, and many of the progeny non-functional. Indeed, HIV pays this cost: a substantial proportion of the HIV in infected cells appear to be so altered by mutation that they are incapable of completing their reproductive cycle (Bagasra *et al.* 1992). But the cost of losing some of the successful parental combinations must not be considered in isolation. Natural selection will weigh this cost against the benefits of divergence from the parental virus. One such benefit is the increased reproduction that results from a temporary escape from immune detection.

Different HIVs probably can recombine genetic instructions (Howell *et al.* 1991), but the potential of this recombination for generating genetic variation is relatively limited; recombination between viruses requires simultaneous infection of a single cell with two different virions. Given that only a small minority of target cells will be infected during most of the time that a person is infected, the probability of simultaneous infection is low. If HIV had a low mutation rate, even these joint infections would not involve much potential for generating variability through recombination. Among HIV then, mutation rate seems to be the primary generator of genetic variability over the short-term time-scale of viral generations.

Why has HIV evolved its two-fold strategy involving both active replication and latency? A consideration of the costs and benefits of replication rates suggests that increased rates of sexual partner change will tend to favour active replication (as opposed to latency), resulting in evolutionary increases in virulence (Ewald 1991, 1994). The distribution of severe and mild HIV infections accords with this view: in regions where sexual partner change appears to be low, the time between infection and AIDS tends to be long (Ewald 1994).

If the hypothesis presented above is correct, HIVs that have not been exposed to high rates of sexual transmission should have lower replication rates and lower mutation rates. A divergent group of HIV (called subtype O) that clusters with HIV-1 should prove particularly useful in assessing whether HIV-1 has recently generated a high mutation rate in response to

high rates of sexual-partner change. The first isolate of this divergent group was called ANT-70 (De Leys *et al.* 1990). According to time scales derived from molecular clocks (Eigen & Nieselt–Struwe 1990), the ANT-70 branch diverged from the other HIV-1 branches centuries ago. Serological surveys in west and westcentral Africa (Gürtler *et al.* 1994; Nkengasong *et al.* 1994) indicate that about 2000 people in the Cameroon and Gabon are infected with viruses that belong to this divergent subtype. No evidence of infection has been found among people to the west of Cameroon (Nkengasong *et al.* 1994). If pandemic HIV-1 generated its mutation-proneness as a result of the high rates of sexual partner change that occurred in central and east Africa during this century (Ewald 1994), subtype O HIVs should have a lower mutation rate.

A similar comparison needs to be made with the other major group of human immunodeficiency viruses: HIV-2, which has been endemic in some geographic areas that have had relatively low rates of sexual partner change. Available evidence indicates that HIV-2s tend to have lower rates of replication and cause more mild infections than pandemic HIV-1s (Ewald 1994). In an *in vitro* assay for mutation rate, reverse transcriptase from HIV-2 generated mutants at a lower rate than reverse transcriptase from HIV-1 (Bakhanashvili & Hizi 1992*a*,*b*). HIV-2's reverse transcriptase might generate mutations more slowly simply because it works more slowly whenever it is synthesizing DNA strands (Hizi *et al.* 1991; Bakhanashvili & Hizi 1992*a*,*b*, 1993), or it might be less mutation-prone because it is less liable to connect mismatched building blocks during the construction of the new strand of DNA (e.g., see Goodman *et al.* 1993). HIV-2s that have been endemic in Senegal should offer a particularly interesting comparison because rates of sexual-partner change there appear to be unusually low, and the progression to AIDS particularly slow (Ewald 1994).

If further tests show that reverse transcription in such HIV-2s and the subtype O HIV-1s are not less mutation-prone, then we can reject the idea that the high mutation rate of pandemic HIV-1 resulted from exposure to high rates of sexual-partner change in central or east Africa. Such evidence would favour the null hypothesis that some other general aspect of lentivirus replication is responsible for the high mutation rate.

3. HUMAN T-CELL LYMPHOTROPIC VIRUSES

The other major group of human retroviruses are the human T-cell lymphotropic viruses (HTLVs), which have a similar genome size but belong to a different subfamily of retroviruses (Temin 1989*a*). HTLVs are geographically widespread and probably have been infecting humans for many millennia (Gessain *et al.* 1991; Goubau *et al.* 1992; Maloney *et al.* 1992). One type, HTLV-I, is globally distributed; prevalences are generally less than 1% but may be above 10% in certain endemic pockets. In about one out of 30 infected people, HTLV-I eventually triggers adult T-cell leukemia/lymphoma (ATL) (Kondo *et al.* 1989; Blattner 1990, Tajima & Ito 1990). Evidence indicates that ATL results from proliferation of infected cells, which in turn is generated by the activity of HTLV inside them (Berneman *et al.* 1992; McGuire *et al.* 1993; Béraud *et al.* 1994; Yoshida 1994).

The substitution rates of HTLV-I are two or more orders of magnitude less than those of HIV-1 (Ratner *et al.* 1991). HTLV's lower rate is largely attributable to a lower reliance on reverse transcriptase for multiplication. Although free HTLV-I virions can infect cells (Fan *et al.* 1992), growth of viral populations inside people occurs mainly by division and fusion of infected cells; similarly, transmission between hosts occurs largely by vertical transfer of infected cells from mother to child (Sugiyama *et al.* 1986; Murphy & Blattner 1988; Yamaguchi 1994). Because this kind of replication and transmission reduces dependence on reverse transcription, it should contribute to a low mutation rate (Temin 1989*b*).

The hypothesis suggested above offers an explanation for HTLV's lower substitution rate. Relying less on production of virions and having relatively less active replication at the beginning of infection, HTLV is less exposed to the immune system. The benefits of generating variation are therefore relatively low. But why, in an ultimate sense, does HTLV rely less on production of virions at the price of reduced replication at the beginning of infection? Consideration of this question requires an analysis of transmission modes and the timing of transmission events.

HTLV-I can be needleborne, sexually transmitted, and transmitted from mother to baby largely through breast milk (Hino *et al.* 1985; Kajiyama *et al.* 1986; Kusuhara *et al.* 1987; Nagamine *et al.* 1991; Kawase *et al.* 1992; Stuver *et al.* 1993). Depending on the relative importances of these transmission modes, one would expect different selective pressures on rates of viral replication and mutation. Maternal transmission of HTLV should place a premium on replication with little stimulation of an immune response because maternally transmitted viruses must survive inside the person from the onset of infection (i.e., in and around the neonatal period) until opportunities for transmission arise (i.e., after the infected baby becomes sexually mature), and then must be transmissible when these opportunities arise. One of the best ways of doing so is by invoking a combination of latency and proviral replication through proliferation of infected cells. If no virions are produced, virions cannot be exposed to the humoral part of the immune system and presentation of viral fragments for control through cellular immunity will be reduced. Indeed, most HTLV infections are characterized by slow development of seroreactivity, extremely low densities of virions, and low to moderate densities of proviral genomes (Agius *et al.* 1988; Wattel *et al.* 1992).

The importance of the different routes of transmission has been most intensively studied in Japan. The geographic clustering of HTLV in Japan indicates that transmission occurs primarily from mother to offspring (Hinuma *et al.* 1982; Tajima *et al.* 1982; Yamaguchi *et al.* 1983; Tajima 1988). Even in cities, HTLV infection has been associated more strongly with the person's geographic origin within Japan than with risky sexual contact (Tajima 1988). This conclusion is supported by a study of HTLV-infected pregnant women: 62% of the women were vertically infected and 23% were sexually infected (Take *et al.* 1993).

This lower rate of sexual transmission probably results partly from attitudes towards sexual contact and birth control. Rates of unprotected sexual contact are relatively low in Japan. Because access to birth control pills is restricted, birth control depends largely on condoms (Trager 1982; Carey *et al.* 1992; Miller *et al.* 1992). A survey completed during the 1970s showed that condoms were used by over 80% of all women using birth control, and over 90% of women in their early twenties; only 3% of women surveyed said that they would use birth control pills (Trager 1982).

Sexual transmission takes place primarily between long-term sexual partners. The slow rise in prevalence among women as a function of age (Tajima *et al.* 1982; Tajima & Ito 1990) suggests that transmission of infections (or the development of detectable evidence of infection) often occurs after many years of sexual contact with an infected partner. This rise in prevalence probably overestimates the degree of sexual transmission because it is inflated by a cohort effect (Ueda *et al.* 1989, 1993). Among a small sample of infected pregnant women, an average of five years elapsed between onset of sexual relations with their husbands and the first seropositive blood sample (Take *et al.* 1993). Probabilities of transmission during sexual contact appear to be about an order of magnitude less for HTLV than for HIV (Blattner 1990).

Maternal transmission also seems to involve low infectivity per instance of contact. Mothers who nurse babies for longer than six months are about three times more likely to infect their babies than mothers who nurse for less than six months (Takahashi *et al.* 1991; Wiktor *et al.* 1993). Seroprevalence among babies stabilizes by about three years of age, suggesting low infectivity and/or slow development of infections within babies (Tajima 1990).

Studies of transmission probabilities indicate that the generally low viral densities reduce the probabilities of infecting a contacted individual: contact with susceptibles leads to infection more frequently as indicators of viral density and activation increase (Sugiyama *et al.* 1986; Hino *et al.* 1987; Sawada *et al.* 1989; Blattner 1990; Ho *et al.* 1991; Scarlatti *et al.* 1991; Stuver *et al.* 1993). Patterns of infection among children of infected parents also indicate that the low viral densities translate into fitness costs for the virus. Children born earlier during a marriage are less likely to be infected, apparently because of low transmissibility from husband to wives or the low transmissibility from mother to baby during the earlier years of a mother's infection (Wiktor *et al.* 1993; Take *et al.* 1994; Umemoto *et al.* 1994).

Taken together, this information supports the following generalizations: HTLV-I infection in Japan is characterized by a high degree of vertical transmission and relatively low potential for sexual transmission. The infrequent opportunities for transmission to susceptible hosts favour immune avoidance through low levels of replication and extremely low levels of virion production, which in turn result in low probabilities of transmission per instance of contact. These characteristics are associated with low mutation rates, which occur through reliance on proliferation of provirally infected cells rather than on reverse transcription.

4. MUTATION RATES AND THE EVOLUTION OF VIRULENCE

The precise reasons for the eventual disintegration of the immune system and the ensuing progression of HIV infection to AIDS are still unclear. Virtually all of the leading hypotheses, however, are based at least partly on high mutation rate. One hypothesis, for example, emphasizes the diversity of variants generated by mutation as the underlying cause of AIDS (Nowak *et al.* 1990). According to another hypothesis, rapidly replicating variants generated by the high mutation rate outcompete the more slowly replicating viruses, and eventually decimate the immune system, bringing on AIDS (Cheng–Mayer *et al.* 1988; Schneweiss *et al.* 1990; Tersmette & Miedma 1990; Gruters *et al.* 1991; Schellekens 1992; Connor *et al.* 1993).

If high mutation rates contribute to the lethality of HIV infections, understanding the evolution of high mutation rates should help us to identify slowly replicating, mutation-averse viruses that might evolve into more dangerous mutation-prone viruses. I suggest that HTLV is such a virus. Transmission of HTLV in populations with high rates of sexual partner change should favour variants that reproduce more rapidly after infecting an individual because the frequent opportunities for sexual transmission would present a lucrative alternative to the infrequent vertical transmission achieved through more cryptic infection. This selection would transform the endemic HTLV from a largely latent or slowly reproducing virus into a virus characterized by both latency and more active replication, like HIV. The hypothesis presented in this paper suggests that once this change in viral characteristics occurs, higher mutation rates will evolve. If high mutation rates, high sexual contact rates, long durations of infections, and infection of CD4$^+$ target cells are largely responsible for the virulence of HIV-1, then another AIDS-like virus could be inadvertently created by favouring the evolution of this collection of characteristics among HTLVs.

The geographic pattern of HTLV disease and sexual transmission suggests that the first steps of this scenario may have already occurred in the Caribbean. In contrast to the situation in Japan, HTLV-I in the Caribbean has the less patchy geographic distribution characteristic of sexual transmission, and it is especially prevalent among people who have many sexual partners (Clark *et al.* 1985; Murphy *et al.*

1989; Rodriguez *et al.* 1993). In Trinidad, for example, it is about six times as prevalent among male homosexuals as among the general population (Bartholomew *et al.* 1987). In contrast to the situation in Japan, condom use is low in most areas of the Caribbean (Schwartz *et al.* 1989; Halsey *et al.* 1992).

Because sexual transmission of retroviruses tends to occur more readily from men to women than from women to men (Padian *et al.* 1991; Stuver *et al.* 1993), sexually transmitted HTLV should be biased toward women. The age at which a female bias in infection rates begins to appear is therefore an indicator of the relative importance of sexually transmitted HTLV. The bias appears shortly after the age of 20 in Jamaica and Barbados, but after the age of 40 in Japan; it continues to rise with age in each of these areas (Tajima *et al.* 1982; Kajiyama *et al.* 1986; Riedel *et al.* 1989; Tajima & Ito 1990; Murphy *et al.* 1991). Only about 7% of the seropositive people in Jamaica are children (Murphy 1990). The analogous figure for Japan is about 17% (Kajiyama *et al.* 1986), this figure underestimates the relative importance of vertical transmission, however, because of a birth cohort effect, which appears to be responsible for much of the higher seropositivity among older age groups in Japan (Ueda *et al.* 1989, 1993) and may have been generated by a reduction in the duration of breast feeding in recent decades (Tajima 1990). Accordingly, the study of HTLV-infected pregnant women (Take *et al.* 1993; see above) suggests that most infections are vertically acquired in Japan. A cohort effect was not found in the Caribbean (Riedel *et al.* 1989).

If the evolutionary scenario proposed above is correct, the HTLVs in the Caribbean should be more harmful than those in Japan. Although conclusive tests of this prediction have not been conducted, the available data do allow a comparison of the rate at which HTLV infections progress to lethal cancers in the two areas. In Japan, people who eventually develop ATL tend to do so late in life, when about 60 years old on average (Tajima 1990; Shimoyama 1991). The absence of ATL among people who did not acquire their infection vertically suggests that ATL in Japan develops primarily and perhaps almost exclusively in infections acquired from parents (Sugiyama *et al.* 1986; Tajima & Ito 1990). The 60-year interval between birth and ATL therefore roughly reflects the time between infection and ATL. In the Caribbean, infected people begin to develop ATL earlier, typically during the early-to-mid-forties (Murphy & Blattner 1988; Murphy *et al.* 1989; Cleghorn *et al.* 1990). Accordingly, the yearly incidence of ATL per HTLV positive adult is about 50% greater in the Caribbean than in Japan (Cleghorn *et al.* 1990). The difference in median age of ATL patients also occurs among people of Japanese and African ethnicity living in the U.S. (the latter mostly of Caribbean extraction) (Levine *et al.* 1994).

These considerations indicate that both sexual partner rate and the virulence of HTLV infections are greater in the Caribbean than in Japan. Additional studies will be needed to determine whether this difference results from differences in the inherent virulence of the HTLVs or from other influences. HTLV researchers have considered a difference in inherent virulence to be improbable without addressing the concordance of such a difference with evolutionary predictions (Levine *et al.* 1994).

Measurements of the error-proneness of HTLV are important for this assessment. If HTLV's low mutation rate per round of replication is attributable entirely to its proviral replication, its reverse transcriptase might be highly error-prone like that of HIV. Direct comparisons of the mutation-proneness of these reverse transcriptases still need to be done. If the reverse transcriptase of HTLV is not less error prone, then the lower mutation rates of HTLV could be interpreted as a non-adaptive side-effect proviral replication. Alternatively, the fitness advantages of proviral replication could be due to both low mutation rate and low immune exposure. Either way, if the low mutation rates of HTLV are attributable solely to a reliance on proviral replication, high rates of sexual partner change could generate high mutation rates by selecting for increased reliance on reverse transcription. This change would be dangerous because high rates of replication and mutation would move HTLV one step closer to the combination of attributes that make HIV-1 so harmful and difficult to control. Alternatively, if HTLV's reduced exposure to an effectively activated immune system favoured a less error-prone reverse transcriptase, increased mutation rates may be more difficult to evolve, leaving a longer window of time before rates of replication and mutation would be coupled.

Data from other geographic areas are more fragmentary, but emerging trends are consistent with the predicted association between sexual partner rates and HTLV virulence (Ewald 1994). The uncertainties associated with the existing evidence emphasize the need for studies that assess rates of mutation, replication, and virulence for the endemic HTLVs of different regions and monitor these characteristics to detect any evolutionary changes. Perhaps the HTLVs in the Caribbean have not yet evolved into a more harmful state because they have not yet evolved increased mutation rates.

HTLV and HIV have similar transmission modes, cell tropisms, genome sizes, and encoded proteins. Although the most dangerous aspects of HTLV pathology concern cancerous growth of white blood cells rather than immune decimation, some ominous similarities exist. In a small minority of HTLV-I infections, for example, some kinds of white blood cells may eventually become decimated. This decimation is more severe as the density of HTLV in the blood increases (Yu *et al.* 1991) and can open the door for opportunistic infections with AIDS-causing organisms such as *Pneumocystis carinii* (Shearer & Clerici 1991).

Until additional data are obtained, it would be prudent to take preventative measures. If the hypothesis presented in this paper is incorrect, investments in reducing rates of sexual transmission would have the well-recognized effect of reducing retroviral prevalence. If the hypothesis is correct such investments will also generate an evolutionary effect, reducing the inherent harmfulness of pathogens like HIV and HTLV, or keeping such pathogens from evolving increased harmfulness in the first place.

I thank W. D. Hamilton for helpful comments on the manuscript. This work was supported by a Faculty Research Award from Amherst College, and a George E. Burch Fellowship of Theoretic Medicine and Affiliated Sciences, which was awarded by the Smithsonian Institution.

REFERENCES

Albert, J., Abrahamsson, B., Nagy, K., Aurelius, E., Gaines, H., Nyström, G. *et al.* 1990 Rapid development of isolate-specific neutralizing antibodies after primary HIV-1 infection and consequent emergence of virus variants which resist neutralization by autologous sera. *AIDS* **4**, 107–112.

Anderson, M.G. & Clements, J.E. 1992 Two strains of SIV (mac) show differential transactivation mediated by sequences in the promoter. *Virology* **191**, 559–568.

Bagasra, O., Hauptman, S.P., Lischner, H.W., Sachs, M. & Pomerantz, R.J. 1992 Detection of human immunodeficiency virus type 1 provirus in mononuclear cells by *in situ* polymerase chain reaction. *N. Engl. J. Med.* **326**, 1385–1391.

Bakhanashvili, M. & Hizi, A. 1992 Fidelity of the reverse transcriptase of human immunodeficiency virus type 2. *FEBS Lett.* **306**, 151–156.

Bakhanashvili, M. & Hizi, A. 1992*a* Fidelity of the RNA-dependent DNA synthesis exhibited by the reverse transcriptases of human immunodeficiency virus type 1 and type 2 and of murine leukemia virus: mispair extension frequencies. *Biochemistry* **31**, 9393–9398.

Bakhanashvili, M. & Hizi, A. 1993*b* The fidelity of the reverse transcriptases of human immunodeficiency viruses and murine leukemia virus, exhibited by the mispair extension frequencies, is sequence dependent and enzyme related. *FEBS Lett.* **319**, 201–205.

Bartholomew, C., Saxinger, W.C., Clark, J.W., Gail, M., Dudgeon, A., Mahabir, B. *et al.* 1987 Transmission of HTLV I and HIV among homosexual men in Trinidad. *J. Am. med. Ass.* **257**, 2604–2626.

Bell, G. 1993 Pathogen evolution within host individuals as a primary cause of sensescence. *Genetica* **91**, 21–34.

Berneman, Z.N., Gartenhaus, R.B., Reitz, M.S., Blattner, W.A., Manns, A., Hanchard, B. *et al.* 1992 Expression of alternatively spliced human T-lymphotropic virus type I pX mRNA in infected cell lines and in primary uncultured cells from patients with adult T-cell leukemia/lymphoma and healthy carriers. *Proc. natn. Acad. Sci. U.S.A.* **89**, 3005–3009.

Blattner, W.A. 1990 Epidemiology of HTLV-I and associated diseases. In *Human retrovirology:* HTLV (ed. W. A. Blattner), pp. 251–265. New York: Raven Press.

Cheng-Mayer, C., Seto, D., Tateno, M. & Levy, J.A. 1988 Biologic features of HIV-1 that correlate with virulence in the host. *Science* **240**, 80–82.

Clark, J., Saxinger, C., Gibbs, W.N., Lofters, W., Lagranade, L., Deceulaer, K. *et al.* 1985 Seroepidemiologic studies of human T-cell leukemia/lymphoma virus type I in Jamaica. *Int. J. Cancer* **36**, 37–41.

Coffin, J.M. 1990 Genetic variation in retroviruses. In *Applied virology research, volume 2. Virus variability, epidemiology and control* (ed. E. Kurstak, R. G. Marusyk, F. A. Murphy & M. J. V. van Regenmortel), pp. 11–31. New York: Plenum.

Connor, R.I., Mohri, H., Cao, Y.Z. & Ho, D.D. 1993 Increased viral burden and cytopathicity correlate temporally with CD4+ T lymphocyte decline and clinical progression in human immunodeficiency virus type 1-infected individuals. *J. Virol.* **67**, 1772–1777.

De Leys, R., Vanderborght, B., Haeseveldt, M.V., Heyndrickx, L., van Geel, A., Wauters, C. *et al.* 1990 Isolation and partial characterization of an unusual human immunodeficiency retrovirus from two persons of westcentral African origin. *J. Virol.* **64**, 1207–1216.

Eigen, M. & Biebricher, C.K. 1988 Sequence space and quasispecies distribution. In *RNA genetics*, vol. 3 (ed. E. Domingo, J. J. Holland & P. Ahlquist), pp. 211–245. Boca Raton, Florida: CRC.

Eigen, M. & Nieselt-Struwe, K. 1990 How old is the immunodeficiency virus? *AIDS* **4** (Suppl. 1), S85–93.

Embretson, J., Zupancic, M., Ribas, J.L., Burke, A., Racz, P., Tenner-Racz, K. et al. 1993 Massive covert infection of helper T lymphocytes and macrophages by HIV during the incubation period of AIDS. *Nature* **362**, 359–362.

Ewald, P.W. 1991 Transmission modes and the evolution of virulence, with special reference to cholera, influenza and AIDS. *Hum. Nature* **2**, 1–30.

Ewald, P.W. 1994 *Evolution of infectious disease.* New York: Oxford University Press.

Fan, N., Gavalchin, J., Paul, B., Wells, K.H., Lane, M.J. & Poiesz, B.J. 1992 Infection of peripheral blood mononuclear cells and cell lines by cell-free human T-cell lymphoma/leukemia virus type I. *J. clin. Microbiol.* **30**, 905–910.

Fauci, A.S., Schnittman, S.M., Poli, G., Koenig, S. & Pantaleo, G. 1991 Immunopathogenic mechanisms in human immunodeficiency virus (HIV) infection. *Ann. intern. Med.* **114**, 678–693.

Gessain, A., Yanagihara, R., Franchini, G., Garruto, R.M., Jenkins, C.L., Ajdukiewicz, A.B. *et al.* 1991 Highly divergent molecular variants of human T-lymphotropic virus type I from isolated populations in Papua New Guinea and the Solomon Islands. *Proc. natn. Acad. Sci. U.S.A.* **88**, 7694–7698.

Goodman, M.F., Creighton, S., Bloom, L.B. & Petruska, J. 1993 Biochemical basis of DNA replication fidelity. *C.r. Biochem. Mol. Biol.* **28**, 83–126.

Goubau, P., Desmyter, J., Ghesquiere, J. & Kasereka, B. 1992 HTLV-II among pygmies. *Nature* **359**, 201.

Gruters, R.A., Terpstra, F.G., Degoede, R.E.Y., Mulder, J.W., DeWolf, F., Schellekens, P.T.A. *et al.* 1991 Immunological and virological markers in individuals progressing from seroconversion to AIDS. *AIDS* **5**, 837–844.

Gürtler, L.G., Hauser, P.H., Eberle, J., Von Brunn, A., Knapp, S., Zekeng, L. *et al.* 1994 A new subtype of human immunodeficiency virus type 1 (MVP-5180) from Cameroon. *J. Virol.* **68**, 1581–1585.

Halsey, N.A., Coberly, J.S., Holt, E., Coreil, J., Kissinger, P., Moulton, L.H. *et al.* 1992 Sexual behavior, smoking, and HIV-1 infection in Haitian women. *J. Am. med. Ass.* **267**, 2062–2066.

Hamilton, W.D. 1980 Sex versus non-sex versus parasite. *Oikos* **35**, 282–290.

Hino, S., Doi, H., Yoshikuni, H., Sugiyama, H., Ishimaru, T., Yamabe, T. *et al.* 1987 HTLV-I carrier mothers with high-titer antibody are at high risk as a source of infection. *Jap. J. Cancer Res.* **78**, 1156–1158.

Hino, S., Yamaguchi, K., Katamine, S., Amagasaki, T., Kinoshita, K., Yoshida, Y. *et al.* 1985 Mother-to-child transmission of human T-cell leukemia virus type-I. *Jap. J. Cancer Res.* **76**, 474–480.

Hizi, A., Tal, R., Shaharabany, M. & Loya, S. 1991 Catalytic properties of the reverse transcriptases of human immunodeficiency viruses type 1 and type 2. *J. biol. Chem.* **266**, 6230–6239.

Ho, G.Y.F., Nomura, A.M.Y., Nelson, K., Lee, H., Polk, B.F. & Blattner, W.A. 1991 Declining seroprevalence and transmission of HTLV-I in Japanese families who immigrated to Hawaii. *Am. J. Epidemiol.* **134**, 981–987.

Holland, J. 1993 Replication error, quaispecies populations, and extreme evolution rates of RNA viruses. In *Emerging viruses* (ed. S. Morse), pp. 203–218. New York: Oxford.

Howell, R.M., Fitzgibbon, J.E., Noe, M., Ren, Z., Gocke, D.J., Schwartzer, T.A. *et al.* 1991 *In vivo* sequence variation of the human immunodeficiency virus type 1 env gene: Evidence for recombination among variants found in a single individual. *AIDS Res. Hum. Retroviruses* **7**, 869–876.

Hübner, A., Kruhoffer, M., Grosse, F. & Krauss, G. 1992 Fidelity of human immunodeficiency virus type I reverse transcriptase in copying natural RNA. *J. molec. Biol.* **223**, 595–600.

Ji, J.P. & Loeb, L.A. 1992 Fidelity of HIV-1 reverse transcriptase copying RNA *in vitro*. *Biochemistry* **31**, 954–958.

Kajiyama, W., Kashiwagi, S., Hayashi, J., Nomura, H., Ikematsu, H. & Okochi, K. 1986 Intrafamilial clustering of anti-ATLA persons. *Am. J. Epidemiol.* **124**, 800–806.

Kawase, K., Katamine, S., Moriuchi, R., Miyamoto, T., Kubota, K., Igarashi, H. *et al.* 1992 Maternal transmission of HTLV-1 other than through breast milk: discrepancy between the polymerase chain reaction positivity of cord blood samples for HTLV-1 and the subsequent seropositivity of individuals. *Jap. J. Cancer Res.* **83**, 968–977.

Kondo, T., Kono, H., Miyamoto, N., Yoshida, R., Toki, H., Matsumoto, I. *et al.* 1989 Age- and sex-specific cumulative rate and risk of ATLL for HTLV-I carriers. *Int. J. Cancer* **43**, 1061–1064.

Kusuhara, K., Sonoda, S., Takahashi, K., Tokugawa, K., Fukushige, J. & Ueda, K. 1987 Mother-to-child transmission of human T-cell leukemia virus type I (HTLV-I): A fifteen-year follow-up study in Okinawa, Japan. *Int. J. Cancer* **40**, 755–757.

Levine, P.H., Manns, A., Jaffe, E.S., Colclough, G., Cavallaro, A., Reddy, G. *et al.* 1994 The effect of ethnic differences on the pattern of HTLV-I-associated T-cell leukemia/lymphoma (HATL) in the United States. *Int. J. Cancer* **56**, 177–181.

Maloney, E.M., Biggar, R.J., Neel, J.V., Taylor, M.E., Hahn, B.H., Shaw, G.M. *et al.* 1992 Endemic human T-cell lymphotropic virus type II infection among isolated Brazilian Amerindians. *J. infect. Dis.* **166**, 100–107.

McGuire, K.L., Curtiss, V.E., Larson, E.L. & Haseltine, W.A. 1993 Influence of human T-cell leukemia virus type I *tax* and *rex* on interleukin-2 gene expression. *J. Virol.* **67**, 1590–1599.

Monk, R.J., Malik, F.G., Stokesberry, D. & Evans, L.H. 1992 Direct determination of the point mutation rate of a murine retrovirus. *J. Virol.* **66**, 3683–3689.

Murphy, E.L. & Blattner, W. 1988 HTLV-I associated leukemia: A model for chronic retroviral diseases. *Ann. Neurol.* **23** (Suppl.), S174–180.

Murphy, E.L., Figueroa, J.P., Gibbs, W.N., Brathwaite, A., Holding-Cobham, M., Waters, D. *et al.* 1989 Sexual transmission of human T-lymphotropic virus type I (HTLV I). *Ann. intern. Med.* **111**, 555–560.

Murphy, E.L., Figueroa, J.P., Gibbs, W.N., Holding-Cobham, M., Cranston, B., Malley, K. *et al.* 1991

Human T-lymphotropic virus type-I (HTLV-I) sero-prevalence in Jamaica. 1. Demographic determinants. *Am. J. Epidemiol.* **133**, 1114–1124.

Nagamine, M., Nakashima, Y., Uemura, S., Takei, H., Toda, T., Maehama, T. *et al.* 1991 DNA amplification of human T Lymphotropic virus type I (HTLV-I) proviral DNA in breast milk of HTLV-I carriers. *J. infect. Dis.* **164**, 1024–1025.

Nkengasong, J.N., Peeters, M., van den Haesevelde, M., Musi, S.S., Willems, B., Ndumbe, P.M. *et al.* 1993 Antigenic evidence of the presence of the aberrant HIV-1 ANT70 virus in Cameroon and Gabon. *AIDS* **7**, 1536–1538.

Nowak, M.A., May, R.M. & Anderson, R.M. 1990 The evolutionary dynamics of HIV-1 quasispecies and the development of immunodeficiency disease. *AIDS* **4**, 1095–1103.

Padian, N., Marquis, L., Francis, D.P., Anderson, R.E., Rutherford, G.W., O'Malley, P.M. *et al.* 1987 Male-to-female transmission of human immunodeficiency virus. *J. Am. med. Ass.* **258**, 788–790.

Piatak, M., Saag, M.S., Yang, L.C., Clark, S.J., Kappes, J.C., Luk, K.C. *et al.* 1993 High levels of HIV-1 in plasma during all stages of infection determined by competitive PCR. *Science* **259**, 1749–1754.

Preston, B.D., Poiesz, B.J. & Loeb, L.A. 1988 Fidelity of HIV-I reverse transcriptase. *Science* **242**, 1168.

Ratner, L., Philpott, T. & Trowbridge, D.B. 1991 Nucleotide sequence analysis of isolates of human T-lymphotropic virus type-1 of diverse geographical origins. *AIDS Res. Hum. Retroviruses* **7**, 923–941.

Riedel, D.A., Evans, A.S., Saxinger, C. & Blattner, W. 1989 A historical study of human T lymphotropic virus type I transmission in Barbados. *J. infect. Dis.* **159**, 603–609.

Roberts, J.D., Bebenek, K. & Kunkel, T.A. 1988 The accuracy of reverse transcriptase from HIV-1. *Science* **242**, 1171–1173.

Rodriguez, E.M., de Moya, E.A., Guerrero, E., Monterroso, E.R., Quinn, T.C., Puello, E. *et al.* 1993 HIV-1 and HTLV-I in sexually transmitted disease clinics in the Dominican Republic. *J. Acquired Immune Defic. Syndr.* **6**, 313–318.

Sawada, T., Tohmatsu, J. & Obara, T. 1989 High risk of mother–to–child transmission of HTLV-I in p40tax antibody-positive mothers. *Jap. J. Cancer Res.* **80**, 506–8.

Scarlatti, G., Lombardi, V., Plebani, A., Principi, N., Vegni, C., Ferraris, G. *et al.* 1991 Polymerase chain reaction, virus isolation and antigen assay in HIV-1-antibody-positive mothers and their children. *AIDS* **5**, 1173–1178.

Schellekens, P.T.A., Tersmette, M., Roos, M.T.L., Keet, R.P., Dewolf, F., Coutinho, R.A. *et al.* 1992 Biphasic rate of CD4+ cell count decline during progression to AIDS correlates with HIV-1 phenotype. *AIDS* **6**, 665–669.

Schneweis K.E., Kleim J.-P., Bailly E., Niese D., Wagner N. & Brackmann H. H. 1990 Graded cytopathogenicity of the human immunodeficiency virus (HIV) in the course of HIV infection. *Med. microbiol. Immunol.* **179**, 193–203.

Schwartz, J.B., Akin, J.S., Guilkey, D.K. & Pagueo, V. 1989 The effect of contraceptive prices on method choice in the Philippines, Jamaica and Thailand. In *Choosing a contraceptive. Method choice in Asia and the United States* (ed. R. A. Bulatao, J. A. Palmore & S. E. Ward), pp. 78–102. Boulder, Colorado: Westview Press.

Shearer, G.M. & Clerici, M. 1991 Early T-helper cell defects in HIV infection. *AIDS* **5**, 245–253.

Shimoyama, M. 1991 Diagnostic criteria and classification of clinical subtypes of adult T-cell leukaemia-lymphoma: A report from the Lymphoma Study Group (1984-7). *Br. med. Haematol.* **79**, 428–437.

Shioiri, S., Tachibana, N., Okayama, A., Ishihara, S.,

Tsuda, K., Essex, M. *et al.* 1993 Analysis of anti-tax antibody of HTLV-I carriers in an endemic area in Japan. *Int. J. Cancer* **53**, 1–4.

Stuver, S.O., Tachibana, N., Okayama, A., Shioiri, S., Tsunetoshi, Y., Tsuda, K. *et al.* 1993 Heterosexual transmission of human T-cell leukemia/lymphoma virus type-I among married couples in southwestern Japan – an initial report from the Miyazaki cohort study. *J. infect. Dis.* **167**, 57–65.

Suárez, P., Valcarcel, J. & Ortin, J. 1992 Heterogeneity of the mutation rates of influenza A viruses: Isolation of mutator mutants. *J. Virol.* **66**, 2491–2494.

Sugiyama, H., Doi, H., Yamaguchi, K., Tsuji, Y., Miyamoto, T. & Hino, S. 1986 Significance of post-natal mother-to-child transmission of human T-lympho-tropic virus type-I on the development of adult T-cell leukemia/lymphoma. *J. med. Virol.* **20**, 253–260.

Tajima, K. 1988 The T- and B-cell malignancy study group. The third nation-wide study on adult T-cell leukemia/lymphoma (ATL) in Japan: Characteristic patterns of HLA antigen and HTLV-I infection in ATL patients and their relatives. *Int. J. Cancer* **41**, 505–512.

Tajima, K. 1990 The T- and B-cell malignancy study group. The fourth nation-wide study on adult T-cell leukemia/lymphoma (ATL) in Japan: estimates of risk of ATL and its geographical and clinical features. *Int. J. Cancer* **45**, 237–243.

Tajima, K. & Ito, S. 1990 Prospective studies of HTLV-I and associated diseases in Japan. In *Human retrovirology: HTLV* (ed. W. A. Blattner), pp. 267–279. New York: Raven Press.

Tajima, K., Tominaga, S., Suchi, T., Kawagoe, T., Komoda, H., Hinuma, Y. *et al.* 1982 Epidemiological analysis of the distribution of antibody to adult T-cell leukemia virus associated antigen (ATLA): Possible horizontal transmission of adult T-cell leukemia virus. *Gann* **73**, 893–901.

Take, H., Umemoto, M., Kusuhara, K. & Kuraya, K. 1993 Transmission routes of HTLV-I: an analysis of 66 families. *Jap. J. Cancer Res.* **84**, 1265–1267.

Takeuchi, Y., Nagumo, T. & Hoshino, H. 1988 Low fidelity of cell-free DNA synthesis by reverse transcriptase of human immunodeficiency virus. *J. Virol.* **62**, 3900–3902.

Temin, H.M. 1989*a* Retrovirus variation and evolution. *Genome* **31**, 17–22.

Temin, H.M. 1989*b* Is HIV unique or merely different? *J. Acquired Immune Defic. Syndr.* **2**, 1–9.

Tersmette, M. & Miedema, F. 1990 Interactions between HIV and the host immune system in the pathogenesis of AIDS. *AIDS* **4** (Suppl. 1), S57–66.

Tremblay M., & Wainberg M. A. 1990 Neutralization of multiple HIV-1 isolates from a single subject by autologous sequential sera. *J. infect. Dis.* **162**, 735–737.

Ueda, K., Kusuhara, K., Tokugawa, K., Miyazaki, C., Hoshida, C., Tokumura, K. *et al.* 1989 Cohort effect on HTLV-I seroprevalence in southern Japan. *Lancet* **2**, 979.

Ueda, K., Kusuhara, K., Tokugawa, K., Miyazaki, C., Okada, K., Maeda, Y. *et al.* 1993 Mother-to-child transmission of human T-lymphotropic virus type I (HTLV-I): an extended follow-up study on children between 18 and 22–24 years old in Okinawa, Japan. *Int. J. Cancer* **53**, 597–600.

Umemoto, M., Take, H., Kusuhara, K. & Kuraya, K. 1994 Risk of HTLV-I infection in Japanese women who are last in birth order. *Cancer Lett.* **76**, 191–195.

Varela-Echavarria, A., Garvey, N., Preston, B.D. & Dougherty, J.P. 1992 Comparison of moloney murine leukemia virus mutation rate with the fidelity of its reverse transcriptase *in vitro*. *J. biol. Chem.* **267**, 24681–24688.

Wattel, E., Mariotti, M., Agis, F., Gordien, E., Le Coeur, F.F., Prin, L. *et al.* 1992 Quantification of HTLV-1 proviral copy number in peripheral blood of symptomless carriers from the French West Indies. *J. Acquired Immune Defic. Syndr.* **5**, 943–946.

Yamaguchi, K., Nishimura, H. & Takatsuki, K. 1983 Clinical features of malignant lymphoma and adult T-cell leukemia in Kumamoto. *Rinsho Ketsekui* **24**, 1271–1276.

Yoshida, M. 1994 Mechanism of transcriptional activation of viral and cellular genes by oncogenic protein of HTLV-I. *Leukemia* (suppl.) **8**, s51–53.

Yu, F., Itoyama, Y., Fujihara, K. & Goto, I. 1991 Natural killer (NK) cells in HTLV-I-associated myelopathy/tropical spastic paraparesis – decrease in NK cell subset populations and activity in HTLV-I seropositive individuals. *J. Neuroimmunol.* **33**, 121–128.

Discussion

C. A. MIMS (*Sheriff House, Ardingly, West Sussex, U.K.*). It has been fascinating to hear Dr Ewald's ideas about the role of viral mutations, viral reactivation from latency, and host behaviour on the evolution of virus diseases. Mutation rates, however, are known to be very great in RNA viruses compared with DNA viruses because of the high frequency of copying errors in RNA polymerases, the final frequency in the virus population depending on growth competition and selective transmission. This is an intrinsic characteristic of RNA viruses.

A second point. Although Dr Ewald suggests that viruses that establish latency and later reactivate are more likely to have high mutation rates, this seems not to be true. Human papillomaviruses, herpesviruses, and polyomaviruses are all DNA viruses and show far less genetic diversity (in so far as this reflects mutation rates) than do RNA viruses such as influenza or rhinoviruses. Indeed, the rate of change in human papillomavirus genomes is so slow that there is less sequence diversity between HPV 16 and HPV 18 than in the HIV virus from a single patient (Ho, L. *et al. J. Virol.* **67**, 6413 (1993); Ong, C.K. *et al. J. Virol.* **67**, 6424 (1993). If points like these were incorporated into Dr Ewald's hypotheses they would be strengthened.

P. W. EWALD. For the reasons that Professor Mims has mentioned, tests of these ideas need to account for the genomes' size and building blocks (DNA or RNA). My presentation focused on retroviruses for these reasons. It is noteworthy, however, that the relative use of DNA or RNA in the life cycle may evolve as a means for altering mutation rate. HTLV, for example, may have evolved a greater reliance on DNA provirus replication to reduce mutation rates whereas Hepatitis B virus may have incorporated an RNA intermediate in its replication cycle to increase mutation rate. The examples that Professor Mims mentions in the second part of his comment are consistent with the general hypothesis that I presented. Low mutation rates are expected among viruses that establish latency and can then be transmitted from the host with little exposure to the immune system. Herpes simplex, for example, does this; after it is activated from latency within neurons, viral progeny can then travel down axons and be transmitted from a blister with little exposure to the immune system. Papillomaviruses too are relatively unexposed to the immune system.

Although influenza viruses do not appear to be transmitted through activation from a latent state, I would expect them to suffer a heavy exposure to the immune system during transmission: those viruses reproducing during the latter part of the infection would benefit by looking different from the viruses that initiated the infection. I would therefore expect them to be toward the mutation-prone end but not as mutation prone as HIV. Studies indicate that they are slightly less mutation prone than HIV. I would expect rhinoviruses to be less mutation-prone than either HIV or influenza, because their exposure to an activated immune system during transmission should be lower, but I am not aware of any direct comparisons to allow a test of this prediction.

One problem with such predictions is that one does not know the fitness tradeoff between the costs and benefits of mutation rates when the mutation rate is near one per genome per replication cycle. It is possible that the extra benefit derived from immune escape after breaking of latency could be swamped by the costs when mutation rates fall into this range. In this case one might find a statistical pile-up of viruses that are exposed to activated immune systems during transmission, with little detectable effect of latency following active replication.

Given that genome size, nucleotide type and other unidentified factors will probably have an effect on mutation rates, I expect that the strongest tests of the general hypothesis will come from studies of closely related viruses that differ in key characteristics such as the breaking of latency in the face of an activated immune system that imposes strong variant-specific culling prior to transmission.

P. J. LACHMANN (*The Royal Society, London, U.K.*). I would like to add three further points to those already made.

1. 'Evolution' of HIV within a patient leading to more rapid growth rate and increased syncytial formation seems also to produce a virus that is less able to be transmitted, possibly because these virulent viruses rapidly lose GP120, so that mutational changes in one host are not necessarily reflected in isolates from the next host.
2. The rate of progression to AIDS is known to be due, at least in part, to both genetic and acquired host factors. Thus sexual promiscuity is related to the incidence of other sexually transmitted disease and thus to the presence of activated T cells in the genital tract. These may be the important factors leading to the accelerated progression of HIV in such subjects.
3. Antibodies to host membrane proteins are effective in lysing SIV grown in xenogenic cells and can presumably limit xenogenic spread of 'promiscuously budding' viruses. It is an interesting possibility that previous alloimmunization (e.g. to HLA) of a wife by her husband may limit infection by the sexual route in the monogamous couple.

The biology of HIV infection is very complex and there are alternative explanations for some of the phenomena you quote in favour of your theory that the rate of sexual partner change determines the mutation rate of the virus.

P. W. EWALD. I agree with Professor Lachman's statements, but none of them negates the hypothesis that I have proposed. We can expect the selective pressures within individuals to favour increased replication rate by a variety of mechanisms. The requirements for transmission to new hosts will tend to select against any that do not also have high capabilities of transmission. Increased syncytial formation appears to be associated with increased replication rate but decreased transmissibility.

The importance of host factors does not negate the

possibility that variation in viral characteristics may be important. HIV-1 and HIV-2 differ in this regard, and available evidence suggests that similar variation occurs among HIV-1s. My point is that the evolutionary associations between replication rate, mutation rate, and social influences on transmission are expected. If we restrict our focus to host factors, we will not recognize and conduct the tests that are needed to determine whether these associations do in fact exist. I agree that alternative hypotheses exist, but we must identify alternative hypotheses before we can test them. The purpose of my presentation was to identify one alternative that needs to be tested.

A. L. HUGHES (*Department of Biology, Pennsylvania State University, U.S.A.*). One of the most fundamental distinctions in evolutionary biology – indeed in all of biology – is the distinction between mutation and substitution. Mutation is an event that happens to a DNA molecule, whereas substitution is a process in a population. When Dr Ewald mentions differences among viruses of similar genome size with respect to 'mutation rate', what sort of evidence is there that these differences are in fact differences in mutation rate rather than in substitution rate?

P. W. EWALD. The difference between HIV-1 and HIV-2 is based on an *in vitro* assay measuring mispair extension frequencies. It therefore is a measure of mutation rate. The lower substitution rate of HTLV relative to HIV must reflect, at least in part, a lower mutation rate per cycle of DNA replication because HTLV DNA replication involves the cell's DNA polymerase, which has a high proof-reading capability. Replication of HIV more frequently involves error prone reverse transcription from an RNA genome. Mutation rate assays of HTLV's reverse transcriptase still need to be done to directly compare HTLV with HIV.

J. D. GILLETT (*London School of Hygiene and Tropical Medicine, U.K.*). What proportion of the population of the monkey, *Cercopithecus aethiops*, is found positive for SIV-1 and SIV-2? I ask this because some 60 years ago the human population in a number of African territories began what we may call a love affair with the hypodermic syringe. Seeing these at that time new fangled instruments used for the improvement of health in various ways, quacks soon came on the scene offering, for suitable payment, injections to cope with every kind of indisposition, real or imagined. The syringes were 'rescued' from hospital and dispensary refuse bins and the like, while the contents varied but were usually diluted with water from the nearby river or swamp.

Now *C. aethiops* is a peri-domestic monkey in many parts of Africa, leaving the nearby forest to raid crop plantations of one sort or another. Men, whose sexual prowess was beginning to wane noticed that the nearby monkeys were, so to speak, always at it. The visiting quacks also noted this and it was not long before they cashed in on the situation and monkey blood, suitably diluted, was on offer for injection. Could this possibly have been the origin of HIV?

P. W. EWALD. Among wild populations of *Cercopithecus aethiops* infected with SIV-2, seroprevalence may range from about 10% to 50%. To my knowledge, no SIV that clusters with HIV-1 has ever been isolated from wild populations of *C. aethiops*. HIV-1 clusters with an SIV from chimpanzees (*Pan troglodytes*) and HIV-2 clusters with an SIV from sooty mangabeys (*Cercocebus atys*). If the invasive practices referred to played a role in the transfer of SIVs to humans they probably would have been transferring the virus from sooty mangabeys or chimps rather than *C. aethiops*. One problem with such syringe scenarios is that they cannot

explain the spectrum of transfers of retroviruses between primate species. They do not, for example, explain the transfer of viruses between *Cercopithecus* species, *Cercocebus* species and mandrills (*Mandrillus sphinx*), nor transfers of HTLV between humans and other primate species that have occurred over the last few millennia. We therefore have evidence that some other route(s) of interspecies transfer is occurring, but no evidence for syringe transfers.

C. E. PARKER (*Zoology Department, University of Oxford, U.K.*). I have two points to make.

First, both vertical and horizontal transmission occur in the Caribbean and in Japan, and both geographical regions show highly clustered, rather than uniformly distributed, prevalence of HTLV-I infection. It is quite erroneous to conclude that: (i) there is a shorter time interval to development of cancer (lymphoma/leukaemia) in the Caribbean; or (ii) to infer, from the misinterpretation, that this is a result of sexually acquired rather than vertically transmitted infection.

Indeed, in all locations where HTLV-I-related lymphoma occurs, this appears to be associated almost entirely with vertically transmitted infection. The requirement, in all geographic areas, for a long time interval for the development of leukaemia reflects multiple, host-dependent factors, as well as the requirement for HTLV-I-related lymphocyte transformation.

Second, in HTLV-I a very brisk immune response is elicited, both humoral, and cell-mediated, and activated cytotoxic T-cells can be detected in freshly isolated lymphocytes from many asymptomatic, HTLV-I-infected people, as well as in rare infected individuals who develop spastic paresis, the other disease associated with HTLV-I. These CTL are chronically activated, and predominantly directed against the tat transactivating protein, implying that chronic expression of HTLV-I tat frequently occurs *in vivo*. This implies that the host immune response, particularly the cytotoxic T-cell response, rather than the particular behavioural characteristics of the infected individual, is the major factor regulating the extent of viral replication.

P. W. EWALD. Both vertical and horizontal transmission do occur in the Caribbean and in Japan. My point is that, according to available evidence, the frequency of vertical relative to sexual transmission is greater in Japan than in Jamaica.

With regard to the time interval to development of cancer, I believe that Dr Parker has misunderstood my argument. The information in the literature shows that ATL tends to occur at younger ages in Jamaica than in Japan. If the frequency of vertical and horizontal transmission were the same in Japan and Jamaica, this difference would indicate that the time between infection and ATL was shorter in Jamaica. In Japan virtually all cases of ATL develop from infections that are acquired during the first few years of life. The median time between infection and ATL is therefore about 60 years in Japan. If the same were true in Jamaica, one would still have a shorter time interval to development of ATL. If, however, more sexually infected individuals develop ATL in Jamaica than in Japan, then we should find a higher proportion of ATL cases among women in Jamaica because sexual transmission occurs mainly from male to female. Although more thorough analyses are needed, published data suggest the latter trend: in Japan about 41% of the ATL patients were females (Tajima *et al.* 1988, Kondo *et al.* 1989), whereas in Jamaica about 57% were females (from table 1 in Murphy *et al.* 1989). If some of this greater ATL rate among Jamaican women is attributable to sexually transmitted HTLV, the time between infection and

ATL would be even shorter in Jamaica than the median age of ATL onset (i.e. 40–45 years), resulting in an even greater difference between Japan and Jamaica in the interval between onset of infection and onset of ATL. These comparisons emphasize the need to directly assess whether the percentage of ATL cases resulting from neonatal infection is as high in Jamaica as it is in Japan.

I think that Dr Parker's second point exaggerates the intensity and effectiveness of the immune response against HTLV. The point concerns primarily the asymptomatic period: by the time ATL sets in, the viruses in those relatively few individuals are stuck in a sinking ship. The immune response to HTLV among asymptomatic HTLV infections is irregular and rather weak; for example, about half are positive for anti-tax antibody (Shioiri *et al. Int. J. Cancer* **53**, 1 (1993)) and about half have induction of HTLV-specific CTLs (Kannagi *et al. Leukemia* **8** (suppl.), s54 (1994)). Dr Parker's data from three asymptomatic patients (Parker *et al. Virology* **188**, 628 (1992)) are consistent with this generalization: one showed no CTL activation and the other two showed mild to moderate activation; antigen specifi-

cities were restricted to tax; and, in her words, the 'magnitude of the anti-tax responses in fresh blood were small'.

The degree of humoral immune response to tax and other viral proteins tends to be positively correlated with transmissibility (Hino *et al.* 1987; Sawada *et al.* 1989; Ho *et al.* 1991; Scarlatti *et al.* 1991; Shioiri *et al.* 1993; Stuver *et al.* 1993). The response therefore appears to be an indicator of viral activity but does not effectively suppress infectivity.

The chronic activation of CTL and its activation against tax do not represent evidence against an evolutionary effect of sexual behaviour on viral replication. According to the arguments I presented, a high mutation rate would be favoured only if the change in protein structure that it generates would allow escape from CTL control, but I know of no evidence suggesting that the irregular CTL response that tends to occur during asymptomatic infection would substantially inhibit sexual or maternal transmission of specific variants that generate this response. Dr Parker too has commented on the lack of any evidence for CTL escape mutants (Parker *et al.* 1992).

8

Gene-for-gene recognition in plant–pathogen interactions

IAN R. CRUTE

Horticulture Research International, Wellesbourne, Warwick CV35 9EF, U.K.

SUMMARY

Mediated through specifically matching allele pairs in the host and pathogen (at resistance and avirulence loci respectively), plants have a refined and highly discriminating capability to recognize and differentiate among genetic variants of potential pathogens. Knowledge of pathogen recognition by plants has primarily resulted from research associated with the selective breeding of crop species for disease resistance. The phenomenon is well described at the cellular, whole plant and population levels in terms of genetics, histology and associated biochemistry. However, a full mechanistic understanding is only now becoming feasible following the isolation and sequencing of putatively interacting plant and pathogen genes.

It is evident that the number of genes in plant genomes capable of specific pathogen recognition is large and that these genes are commonly clustered in complex loci sometimes comprising genes involved in the recognition of more than one taxonomically unrelated pathogen. For some genes, alleles with different recognition capabilities have been identified while there is also evidence for the existence of genes expressing identical recognition capability present at different loci in the same species as well as in different plant species. It seems likely that resistance genes are members of substantial multi-gene families and there is evidence that novel recognition capability may be created through interallelic recombination events.

In natural populations, there can be a spatial mosaic of resistance genotypes with individuals carrying a range of resistance alleles at a varying number of loci, some of which will exert selection on the pathogen population only under certain conditions and at certain stages of development. There have been few studies of the costs and benefits of particular resistance genes in the presence or absence of pathogen variants against which a gene may be effective or ineffective. The stage of host development at which the pathogen exerts the greatest impact on survival and fecundity is of particular relevance in this respect, as is the influence of mother plant resistance on the characteristics and quantity of inoculum to which progeny are exposed.

1. INTRODUCTION

The existence of plant genes providing resistance to pathogens was demonstrated soon after the rediscovery of Mendel's seminal studies on inheritance. Biffen (1905) demonstrated that a single locus was responsible for the resistance of some wheat cultivars to yellow rust (*Puccinia striiformis*). Since this early work, many hundreds of genes associated with resistance to a diversity of pathogens have been identified in numerous plant species, primarily those of agricultural significance. At an early stage, it was discovered that genes at different loci could be responsible for resistance to different pathogenic variants (pathotypes) (McRostie 1919) but the full significance of this observation only became evident after the gene-for-gene relationship was elucidated by Flor (1956, 1971) in the course of 40 years of research on the interaction between flax (*Linum ultissimum*) and the rust, *Melampsora lini*.

It became evident, following Flor's classical work, that for many host–parasite relationships, matching gene-pairs (resistance and avirulence genes respectively) controlled the outcome of different genotype combinations (Crute 1985). This paper provides a short and selective discussion of information about genotype-specific recognition of pathogens by plants within the context of genome organization, phenotypic expression, natural plant population structure and evolution.

2. THE GENE-FOR-GENE RELATIONSHIP

In gene-for-gene relationships, the expression of resistance or susceptibility of the host to a particular pathogen is conditional on the pathogen genotype and the observed virulence of the pathogen is conditional on the host genotype. Specifically matching gene-pairs determine the outcome of any particular genotype × genotype interaction. Compatibility (uninhibited

Table 1. *Features of a hypothetical gene-for-gene relationship involving three interacting gene pairs*

(3 represents complete compatibility (susceptibility/virulence) and 0 represents the highest level of incompatibility. Gene pairs have an epistatic relationship such that $R1/A1$ is epistatic to $R2/A2$ which is epistatic to $R3/A3$.)

			\multicolumn{8}{c}{pathogen genotypes}
			A1	a1	A1	A1	a1	a1	A1	a1
			A2	A2	a2	A2	a2	A2	a2	a2
			A3	A3	A3	a3	A3	a3	a3	a3
host genotypes										
R1	R2	R3	0	1	0	0	2	1	0	3
r1	R2	R3	1	1	2	1	2	1	3	3
R1	r2	R3	0	2	0	0	2	3	0	3
R1	R2	r3	0	1	0	0	3	1	0	3
r1	r2	R3	2	2	2	3	2	3	3	3
r1	R2	r3	1	1	3	1	3	1	3	3
R1	r2	r3	0	3	0	0	3	3	0	3
r1	r2	r3	3	3	3	3	3	3	3	3

pathogen development and reproduction in the absence of an effective host defence response) is the outcome of a host x pathogen combination unless an allele for resistance at a particular host locus is specifically matched by an allele for avirulence at a particular pathogen locus; under these circumstances, degrees of incompatibility (reduced pathogen development and reproduction associated with an effective host defence response) are expressed depending on the particular matching gene-pair. Gene pairs resulting in incompatibility are epistatic over gene pairs that would otherwise result in compatibility. Gene pairs conditioning higher degrees of incompatibility are epistatic over gene pairs associated with lower degrees of incompatibility, although phenotypic variation indicative of genetic additivity has also been reported when more than one gene pair conditioning incompatibility is effective. Table 1 illustrates the features of a hypothetical gene-for-gene relationship involving three epistatic matching gene pairs when 3 represents complete compatibility and 0 represents the highest degree of incompatibility. Assuming that genes controlling the interaction are at separate loci, and ignoring heterozygotes, there are 2^n possible host and pathogen genotypes and $(2^n)^2$ unique genotype x genotype combinations (where n is the number of matching gene pairs).

It appears that there is a large number of genes in a plant species capable of functioning in the genotype-specific recognition of pathogens. For example, in wheat, a total of more than 90 genes have been identified which condition isolate specific resistance to three rust species (*Puccinia striiformis*, *P. recondita* and *P. graminis*) and powdery mildew (*Erysiphe graminis*). Only one of these genes (*Lr20* and *Sr15*) is thought to be involved in the recognition of more than one pathogen (see Crute 1985 for references).

The suggestion that gene-for-gene specificity is in some way an artifact of cultivation are not substantiated by an increasing number of investigations of natural plant pathosystems. In fact, the narrow genetic

base of some crop species may mean that they are relatively impoverished with respect to genes for pathogen recognition; the need to exploit additional genetic diversity among wild progenitors is a familiar approach for plant breeders seeking to improve disease resistance. Studies on two ruderal weed species (*Senecio vulgaris* and *Arabidopsis thaliana*) have readily demonstrated the existence of a substantial number of genes identified by their ability to discriminate among a relatively restricted sample of pathogen isolates (*Erysiphe fischeri* and *Peronospora parasitica* for the two host species, respectively) (Harry and Clarke 1986; Bevan *et al.* 1993a; Holub *et al.* 1994).

The gene-for-gene relationship provides the basis for the discrimination of pathogenic variants (pathotypes) by their virulence or avirulence on particular genotypes of a host species. However, in addition to this sub-specific variation among genotypes within species, pathotypic variation is also frequently evident at the level of host species, genus or family. For example, the downy mildew pathogen, *Peronospora parasitica*, is only . pathogenic on cruciferae but, although it has been recorded on over 50 host genera, pathotypes are restricted to the host genus or species of origin. Hence, isolates from *Arabidopsis thaliana* are avirulent on *Brassica* spp. and vice versa. Pathotypes of pathogens discriminated by their specific adaptation to higher host taxa are often formally ascribed to *formae speciales*. It is clear that such specificity represents the outcome of a co-evolved parasitic symbiosis and while there is some evidence that this level of specificity may also be a reflection of gene-for-gene recognition the subject continues to be debated (Heath 1987, 1991; Newton & Crute 1989; Tosa 1989, 1992).

3. THE ORGANIZATION AND STRUCTURE OF PATHOGEN RECOGNITION LOCI

Studies on the inheritance of isolate-specific resistance to flax rust by Flor and others have identified the existence of genes expressing at least 32 different specificities and organized in five linkage groups (*K*, *L*, *M*, *N* and *P*) (see Islam & Shepherd 1991a for a recent review). While the existence of separate loci within linkage groups *K*, *M*, *N* and *P* have been demonstrated, the 14 specificities at the *L* locus appear to be allelic. Studies have been made of progeny from test crosses between a homozygous susceptible genotype and 27 of the possible 91 *L* group heterozygotes. Among large numbers of individuals, rare susceptible plants were located as were plants expressing non-parental resistance phenotypes (so-called 'modified recombinants'); however, no individuals were found which expressed the combined specificity of both *L* group parents. Some susceptible individuals yielded resistant revertants on selfing and plants expressing novel specificity to nine different rust pathotypes were identified among progeny of revertants, susceptible recombinants and modified recombinants. Interallelic recombination, providing the mechanism for the generation of novel recognition capability, has been postulated to explain these data

(Islam *et al.* 1989, 1991; Islam & Shepherd 1991). Although there are few other well-authenticated examples of allelism at parasite recognition loci in plants, it is entirely feasible that variants of susceptibility alleles (that would be impossible to discriminate phenotypically) could also play a part in the generation of novel specificity through rare interallelic recombination events.

A more common circumstance than allelism appears to be the clustering of genes capable of specific parasite recognition within large, complex loci. Saxena & Hooker (1968) conducted a classical study of genes in maize for resistance to the rust fungus, *Puccinia sorghi* located on the short-arm of chromosome 10. Sixteen recognition genes (*Rp5*, *Rp6* and 14 genes previously thought to be alleles of *Rp1* – *A* to *N*) map within this region, the existence of separate linked loci being demonstrated by recombination events and predicted patterns of resistance to diagnostic isolates of the pathogen. More recent studies of loci at *Rp1* have indicated meiotic instability observed as the unexpected appearance of susceptible individuals in test-cross progeny between parents homozygous susceptible and resistant at *Rp1* loci (Hulbert & Bennetzen 1991; Hong *et al.* 1993; Sudupak *et al.* 1993). This phenomenon has been shown to result from unequal crossing-over as indicated by non-parental flanking RFLP loci. Mispairing of sequence duplications provides a mechanism leading to gene duplication and associated elimination. It is probable that the necessary repeats to facilitate mispairing are provided in *Rp1* regions which represent a multigene family of duplicated homologues distinguished by their diagnostic pathogen recognition capability. Additional evidence for the occurrence of repetitive sequences in the *Rp1* region was provided by the discovery of two genomic clones which identify a variable number of loci tightly linked to *Rp1* in different maize lines.

In lettuce (*Lactuca sativa*), identified linkage groups contain genes for recognition of several unrelated parasites. One linkage group contains a gene for resistance to an aphid species (*Pemphigus bursarius*) in addition to eight genes for specific resistance to downy mildew (*Bremia lactucae*). Two further downy mildew resistance genes occur in a linkage group along with a gene for resistance to turnip mosaic virus and a gene for resistance to the root pathogen, *Plasmopara lactucae–radicis* (Landry *et al.* 1987; Kesseli *et al.* 1993). Genes in tomato (*Lycopersicon esculentum*) for resistance to root-knot nematode (*Meloidogyne* spp.) and leaf mould (*Cladosporium fulvum*) are also linked (Dickinson *et al.* 1993; Jones *et al.* 1993).

There are probably several selective advantages contributed by the common existence of complex loci such as those described above. Such arrangements allow multiple parasite recognition capabilities to be assembled and retained in a single haplotype while out-breeding and recombination can readily generate new combinations of specificities. Mispairing, intragenic recombination and gene duplication may facilitate the evolution of novel recognition capability.

In contrast to the occurrence of complex loci comprising several genes with distinct recognition capabilities, there is recent data also indicating the existence of functionally identical duplicate genes at independent loci (Teverson *et al.* 1994). In the interaction between *Phaseolus* beans and the bacterial pathogen, *Pseudomonas syringae* pv. *phaseolicola*, digenic and trigenic segregation ratios have been observed among F_2 progeny from crosses between cultivars previously demonstrated to carry functionally identical recognition genes as indicated by their reaction to transconjugant strains of the bacterium carrying single cloned avirulence genes (*avrPphB1.R3* and *avrPphC1.R1*). (All notations for bacterial avirulence genes follow the proposals of Vivian and Mansfield (1993)). Intriguingly, one of the genes apparently occurring at duplicate loci (*R3*) appears to be identical or tightly linked to the so-called *I* gene in bean which provides isolate specific resistance to bean common mosaic virus and several other related potyviruses (J. D. Taylor, personal communication).

The existence of genes in different plant species with seemingly identical recognition capability has also now been firmly established. Sequencing of two avirulence genes, isolated respectively from the crucifer pathogen, *Pseudomonas syringae* pv. *maculicola* (*avrPmaA1.RPM1*) and the pea pathogen, *P. syringae* pv *pisi* (*avrPpiA1.R2*), proved them to be near-identical. In transconjugant strains of the two pathovars these avirulence genes also exhibited identical specificity of interaction with the respective host genes: *R2* (from pea) and *RPM1* (from *Arabidopsis thaliana*) indicating functional homology (Dangl *et al.* 1992). Furthermore, in bean, recognition genes at two independent loci were identified by a transconjugant of the bean pathogen *Pseudomonas syringae* pv. *phaseolicola* carrying *avrPpiA1.R2* (Fillingham *et al.* 1992).

Martin *et al.* (1993) have recently reported the cloning and sequencing of a gene from tomato (*Pto*) which provides specific resistance to isolates of *Pseudomonas syringae* pv. *tomato* carrying the matching avirulence gene, *avrPto.C1*. It has been demonstrated that at least six homologous expressed genes are clustered at the *Pto* locus and homologues of *Pto* occur in all 11 other plant species examined. Somewhat unexpectedly, the *Pto* protein has similarity with serine–threonine protein kinases and has no obvious membrane-spanning domain suggesting a signal transduction rather than a receptor function.

On the basis of all the evidence so far available, it would not be surprising if plant genes, identified by their specific pathogen-recognition capability, frequently prove to be members of substantial, linked, multi-gene families well-conserved between plant families (Pryor 1987).

4. NATURAL PATHOSYSTEMS

Most of what has been learnt about the genetics of plant–pathogen interactions results from investigations of crop pathosystems but such analyses must inevitably reflect their man-guided and recent evolutionary history. By contrast with their progenitors, most crops probably have a relatively narrow genetic base. For example, horticultural forms of *Brassica oleracea* (cabbage, cauliflower, broccoli and Brussels sprout) are unusual phenotypic variants likely to have originated once and thereby harbouring relatively little genetic diversity. Most crops grow in stable or even optimized environments with a sufficiency of nutrients and water and minimization of competition with other species (weeds); this represents a marked contrast with natural plant populations. While the sizes of natural plant populations may fluctuate widely over time, particularly in the case of ruderal annual species, the sizes of crop populations remain relatively stable under the influence of frequent replacement by man. Given these and other contrasts between crop and natural pathosystems, it seems probable that artificial selection for improved disease resistance in crops could favour a restricted sub-set of the genes naturally selected to defend plants from pathogens.

In addition to the large numbers of specific recognition genes that have been identified in some wild plant species, natural populations also exhibit considerable spatial diversity with respect to these genes. A study of resistance in the annual composite weed, groundsel (*Senecio vulgaris*) to the powdery mildew pathogen, *Erysiphe fischeri*, revealed ten different patterns of phenotypic response to five pathogen isolates among a sample of 75 plants from $1\,m^2$ (Bevan *et al.* 1993*b*) and little additional variation was evident in a sample derived from $6\,km^2$.

The characteristics commonly associated with resistance genes selected in crop breeding programmes include: dominance and a high level of expressivity, often resulting in the complete absence of pathogen reproduction, at all developmental stages, in all tissues and over a range of environments. There are exceptions to these generalizations and temperature-sensitive genes or those which exert their effect only in adult plants are not uncommon in crop species (reviewed by Crute 1985). There are, however, some indications from recent studies with non-crop species that resistance genes conforming to the general pattern of those common in crops do not necessarily predominate. Genes with relatively low expressivity appear to be commonly located in *Arabidopsis thaliana* for response to different isolates of *Peronospora parasitica* (downy mildew) and in *Senecio vulgaris* for response to different isolates of *Erysiphe fischeri* (Bevan *et al.* 1993*a*; Holub *et al.* 1994). Furthermore, in the former case, genes whose expression is associated with extensive necrosis of the mesophyll in the absence of fungal colonization (referred to as 'pitting') have also been observed (Holub *et al.* 1994); such phenotypic expression of resistance is not commonplace in crops and would probably be consciously rejected in the process of selection. In *Senecio vulgaris*, resistance that was only expressed in seedlings or only in adult plants or only at certain temperatures was also observed (Bevan *et al.* 1993*c*).

5. DISCUSSION

A substantial number of genes occur in plant species which allow the genotype-specific recognition of potential pathogens; the actual numbers for any particular individual or species are unknown, their identification being dependent on diagnostic pathogen variants. These genes occur within plant genomes in distributed clusters reminiscent of multi-gene families and considerable variation occurs in the phenotypic expression of incompatibility dependent on the particular interacting gene-pair. The common occurrence of these genes, at least in herbaceous annuals and deciduous perennials, suggests that they are selectively advantageous even though pathotypes capable of rendering the resistance they confer ineffective are frequently encountered.

There is likely to be a cost to the individual of carrying any particular recognition gene both in the presence and the absence of the pathogen. Similarly, in the presence of the pathogen, there is a benefit of carrying an effective recognition gene and a cost of not doing so. The benefit might be considered to be greatest at the stage of plant development or under those environmental conditions when the pathogen would be expected to exert its greatest effect on host survival and fecundity. Critical studies are required to examine these costs and benefits and to determine when and under what circumstances pathogens are likely to exert strong selection.

The progeny of inbreeding host species in the proximity of a diseased mother plant are likely to be particularly vulnerable to pathogenesis due to their genetic identity. Hence, genes which are effective and exert selection on the pathogen only in adult plants would reduce levels of inoculum to which establishing seedlings might be exposed. Alternatively or in addition, the occurrence among progeny of rare recombinant variants, expressing novel pathogen recognition capability, would allow survival of the potentially well-adapted parental genotype otherwise threatened by a pathotype with matching virulence. While outbreeding allows for the regular reassortment of genes involved in pathogen recognition, genetic mechanisms that facilitate the creation of new pathogen recognition capability may be an essential prerequisite for successful adoption of an inbreeding reproductive strategy.

An important feature of competitiveness in relation to the successful colonization of a new habitat by immigrants is likely to be the possession of a novel and more effective specific pathogen recognition capability in comparison with previously established individuals already subject to pathogenesis by matching pathotypes.

Although crop cultivars, in common with other plants, may carry many genes for specific recognition of pathogens, they have frequently been selected for genetic uniformity; examples of extreme uniformity are provided by some vegetatively propagated crops and F_1 hybrid cultivars of seed-propagated crops. The same genotypes may also be grown over prolonged periods of time. The greater prominence of diseases in crops compared to plants in natural plant populations may be a direct consequence of this circumstance. In contrast, natural plant populations comprise a dynamic temporal and spatial mosaic of genotypes carrying variable numbers of different pathogen-recognition genes with variable expressivity. Some of these genes will exert selection on the pathogen population only under certain environmental conditions or at certain stages of host development.

In any diverse population of plants, at any time, some will be selectively advantaged by virtue of the pathogen recognition genes they carry; selection on the pathogen population will ensure that at a different time selection favours different host individuals. In this way, the overall influence of pathogenesis may be minimized with recognition genes enjoying renewed cycles of effectiveness as discontinuities in epidemics and cycles of infection are experienced through the annual and deciduous habit or diverse ecological influences on population stability and size.

REFERENCES

Bevan, J.R., Crute, I.R. & Clarke, D.D. 1993*a* Variation for virulence in *Erysiphe fischeri* from *Senecio vulgaris*. *Pl. Path.* **42**, 622–635.

Bevan, J.R., Clarke, D.D. & Crute, I.R. 1993*b* Resistance to *Erysiphe fischeri* in two populations of *Senecio vulgaris*. *Pl. Path.* **42**, 636–646.

Bevan, J.R., Crute, I.R. & Clarke, D.D. 1993*c* Diversity and variation in expression of resistance to *Erysiphe fischeri* in Senecio vulgaris. *Pl. Path.* **42**, 647–653.

Biffen, R.H. 1905 Mendel's laws of inheritance and wheat breeding. *J. agric. Sci.* **1**, 4–48.

Burdon, J.J. 1987 *Diseases and plant population biology*, 208 pp. Cambridge: Cambridge University Press.

Crute, I.R. 1985 The genetic bases of relationships between microbial parasites and their hosts. In *Mechanisms of resistance to plant disease* (ed. R. S. S. Fraser), pp. 80–142. Dordrecht, Martinus Nijhoff and W. Junk.

Dangl, J.L., Ritter, C., Gibbon, M.J. *et al.* 1992 Functional homologues of the *Arabidopsis RPM1* disease resistance gene in bean and pea. *Pl. Cell* **4**, 1359–1369.

Dickinson, M.J., Jones, D.A. & Jones, J.D.G. 1993 Close linkage between the *Cf-2/Cf-5* and *Mi* resistance loci in tomato. *Molec. Pl. Micro. Interact.* **6**, 341–347.

Fillingham, A.J., Wood, J, Bevan, J.R. *et al.* 1992 Avirulence genes from *Pseudomonas syringae* pathovars *phaseolicola* and *pisi* confer specificity towards both host and non-host species. *Phys. Mol. Pl. Path.* **40**, 1–15.

Flor, H.H. 1956 The complementary genetic systems in flax and flax rust. *Adv. Genet.* **8**, 29–54.

Flor, H.H. 1971 The current status of the gene-for-gene concept. *A. Rev. Phytopathol.* **9**, 275–296.

Heath, M.C. 1987 Evolution of plant resistance and susceptibility to fungal invaders. *Can. J. Pl. Pathol.* **9**, 389–397.

Heath, M.C. 1991 The role of gene-for-gene interactions in the determination of host species specificity. *Phytopathology* **81**, 127–130.

Holub, E.B., Beynon, J.L. and Crute, I.R. 1994 Phenotypic and genotypic characterization of interactions between isolates of *Peronospora parasitica* and accessions of *Arabidopsis thaliana*. *Molec. Pl. Micro. Interact.* **7**, 223–239.

Hong, K.S., Richter, T.E., Bennetzen, J.L. & Hulbert, S.H. 1993 Complex duplications in maize lines. *Molec. gen. Genet.* **239**, 115–121.

Hulbert, S.H. & Bennetzen, J.L. 1991 Recombination at the *Rp1* locus of maize. *Molec. gen. Genet.* **226**, 377–382.

Islam, M.R. & Shepherd, K.W. 1991*a* Present status of genetics of rust resistance in flax. *Euphytica* **55**, 255–267.

Islam, M.R. & Shepherd, K.W. 1991*b* Analyses of phenotypes of recombinants and revertants from test-cross progenies involving genes at the *L* group, conferring resistance to rust in flax. *Hereditas* **114**, 125–129.

Islam, M.R., Shepherd, K.W. & Mayo, G.M.E. 1989 Recombination among genes at the *L* group in flax conferring resistance to rust. *Theor. appl. Genet.* **77**, 540–546.

Islam, M.R., Shepherd, K.W. & Mayo, G.M.E. 1991 An analysis of reversion among recombinants involving genes of the *L* group, conferring resistance to rust in flax. *J. Genet. & Breed.* **45**, 181–188.

Jones, D.A., Dickinson, M.J., Balint-Kurti, P.J., Dixon, M.S. & Jones, J.D.J. 1993 Two complex resistance loci revealed in tomato by classical and RFLP mapping of the *Cf-2*, *Cf-4*, *Cf-5* and *Cf-9* genes for resistance to *Cladosporium fulvum*. *Molec. Pl. Micro. Interact.* **6**, 348–357.

Kesseli, R., Witsenboer, H., Stanghellini, M., Vandermark, G. & Michelmore, R.W. 1993 Recessive resistance to *Plasmopara lactucaeradicis* maps by bulked segregant analysis to a cluster of dominant resistance genes in lettuce. *Molec. Pl. Micro. Interact.* **6**, 722–728.

Landry, B.S., Kesseli, R., Farrara, B. & Michelmore, R.W. 1987 A genetic map of lettuce (*Lactuca sativa* L.) with restriction fragment length polymorphism, isozyme, disease resistance and morphological markers. *Genetics* **116**, 331–337.

Martin, G.B., Brommonschenkel, S.H., Chungwongse, J. *et al.* 1993 Map-based cloning of a protein kinase gene conferring disease resistance in tomato. *Science* **262**, 1432–1436.

McRostie, G.P. 1919 Inheritance of anthracnose resistance as indicated by a cross between a resistant and a susceptible bean. *Phytopathology* **9**, 139–148.

Newton, A.C. & Crute, I.R. 1989 A consideration of the genetic control of species specificity in fungal plant pathogens and its relevance to a comprehension of the underlying mechanisms. *Biol. Rev.* **64**, 35–50.

Pryor, T. 1987 The origin and structure of fungal disease resistance genes in plants. *TIG* **3**, 157–161.

Tosa, Y. 1992 A model for the evolution of *formae speciales* and races. *Phytopathology* **82**, 728–730.

Saxena, K.M.S. & Hooker, A.L. 1968 On the structure of a gene for disease resistance in maize. *Proc. natn. Acad. Sci. U.S.A.* **61**, 1300–1305.

Sudupak, M.A., Bennetzen, J.L. & Hulbert, S.H. 1993 Unequal exchange and meiotic instability of disease resistance genes in the *Rp1* region of maize. *Genetics* **133**, 119–125.

Teverson, D.M., Taylor, J.D., Crute, I.R. *et al.* 1994 Pathogenic variation in *Pseudomonas syringae* pv. *phaseolicola* III. Analysis of the gene-for-gene relationship between pathogen races and *Phaseolus vulgaris* cultivars. *Pl. Path.* (In the press.)

Vivian, A. & Mansfield, J. 1993 A proposal for a uniform genetic nomenclature for avirulence genes in phytopathogenic pseudomonads. *Molec. Pl. Micro. Interact.* **6**, 9–10.

9

MHC polymorphism and parasites

JAN KLEIN[1,2] AND COLM O'HUIGIN[1]

[1] *Max-Planck-Institut für Biologie, Abteilung Immungenetik, Corrensstr. 42, D-72076 Tübingen, Germany*
[2] *Department of Microbiology and Immunology, University of Miami School of Medicine, Miami, Florida 33136, U.S.A.*

SUMMARY

The major histocompatibility complex (MHC) polymorphism is marked by the existence of allelic lineages that are extremely old, having been passed from one species to another in an evolutionary line of descent. Each species has several of these lineages and many of their more recent derivatives, the actual alleles. The lineages are separated by large genetic distances and are characterized by the presence of short sequence motifs which, at the protein level, have remained virtually unaltered for over 40 million years. Several explanations for the MHC polymorphism have been proposed. We argue that the only one consistent with the entire body of knowledge about the MHC is an explanation based on the immune response to parasites. Furthermore, we propose that parasites coevolving with their hosts have had a major influence on MHC polymorphism, whereas parasites that switched hosts recently and became very virulent have had little effect. The latter category includes micro- and macroparasites responsible for the major human infectious diseases. This hypothesis explains why no convincing association between human leucocyte antigen (*HLA*) alleles and resistance to infectious disease can thus far be documented, and indicates the direction in which the search for such associations should be taken.

1. SELF–NONSELF DISCRIMINATION

Vertebrates, for reasons that are not at all obvious, perhaps because of the growing size and complexity of their bodies (Klein 1989), embarked on an evolutionary pathway leading to the emergence of an entirely new defence system aimed at protecting them against both micro- and macroparasites. The essential feature of this system is that it anticipates every possible molecule the parasites might come up with. A fundamental predicament in designing it is how to distinguish non-self from self. All living creatures, parasites and their hosts included, are constructed from the same building blocks assembled in a wide variety of permutations. Limiting our considerations to proteins alone, we note that all organisms use the same 20 or so amino acids to construct unique polypeptides, which differ in sequence between parasites and their hosts. Unfortunately, it is apparently impossible to design a receptor capable of recognizing a protein molecule as a whole and thus differentiating it from all other protein molecules. Recognition is always limited to a small portion of the ligand: a peptide, a three-dimensional motif, or a combination of both. And here are the horns of the dilemma: because there is a considerable overlap between the peptides of the parasite and those of its host, the host must find a way of ignoring the overlapping peptides (otherwise the defence mechanisms would turn against the host itself) and focusing only on those unique to the parasite.

Natural selection has found a solution to this dilemma so simple and so ingenious that even the smartest speculative biologists were unable to figure it out before experimentalists discovered how it works. It requires three molecules acting in concert: the major histocompatibility complex (MHC), the T-cell receptor (TCR), and the immunoglobulin (Ig) molecules. The MHC molecule acts first. It focuses exclusively on proteins and ignores carbohydrates, lipids, nucleic acids, and other organic substances. Furthermore, it deals best with protein fragments, peptides, although it can also handle entire protein molecules. It is a receptor for peptides, but a receptor of very limited specificity. It exists in two variants, class I and class II (Klein 1986), each of which has somewhat different restrictions on peptide binding (Rammensee *et al.* 1993; Germain 1994). The peptide-binding groove of the class I molecule is closed so that only peptides of a certain length (8–10 amino acid residues) fit into it and are bound by their termini. It has pockets into which side chains of the peptides, anchor residues, must fit for optimal binding. The anchors and the fixed distances between them determine most of the specificity of the peptide binding which thus remains very broad, allowing thousands of different peptides to be accommodated by a particular MHC molecule. The groove of the class II molecule, on the other hand, is open and accommodates peptides that extend beyond its edges (length range of 12–24 residues); the peptides are not bound by their termini. The specificity of the binding is achieved through the interaction between stretches of amino acid residues in the peptide and in the groove, whereby the side chains of the former must again fit into the pockets of the latter.

Peptide binding occurs when the dimeric MHC molecules are assembled from their constituent chains and the complex is then displayed on the surface of the 'antigen presenting cells' (Neefjes & Momburg 1993). The MHC molecules bind self and non-self peptides indiscriminately, most of the time the former. The discrimination is then accomplished by a neat trick during the maturation of the immune system. The peptide–MHC protein combination is recognized by the TCR and this recognition is highly specific. For each MHC-peptide complex combination, there is a T lymphocyte clone that recognizes this and virtually no other combination. Initially, there are T lymphocyte clones in the developing immune system for *any* combination of peptide and MHC protein, both self and non-self. At a later stage, when there are still very few non-self peptides in the body, all lymphocytes reactive with self peptide–MHC protein complexes are purged from the immune system, which is then left with cells capable of recognizing non-self peptide–MHC protein complexes only (Miller 1992). When such complexes occur in the body during a parasitic infection, the responding clone is stimulated and either becomes involved in the attack on the invader itself or releases mediators that stimulate B lymphocytes to secrete Ig molecules, the 'antibodies' of the classical immune response.

The amino acid residues of the groove that come into contact with the peptide, the peptide-binding region or PBR, are highly variable and the corresponding part of the specifying gene is highly polymorphic (Hughes & Nei 1989). We will illustrate the features of this polymorphism by taking one of the primate class II loci, the *DRB1* locus, as an example. The number of alleles described at the *DRB1* locus in the human species alone is edging toward 100 (Marsh & Bodmer 1993). Several hundred more have been reported for a variety of other primate species (O'hUigin *et al.* 1993), none of which, however, has been studied as thoroughly as *Homo sapiens*.

2. ALLELIC LINEAGES AND MOTIFS

A conspicuous characteristic of *DRB1* polymorphism is that the alleles fall into groups that we shall refer to as *allelic lineages*. Each lineage is distinguished by one or more diagnostic *motifs*: short stretches of nucleotide sequences with substitutions peculiar to that lineage. The motifs are localized in (non-contiguous) segments of the sequence specifying the PBR. Because the substitutions are primarily non-synonymous, the motifs can be described most succinctly when translated into amino acid sequences. Hence, when we speak of 'allelic lineages' but use amino acid notation for their motifs, it is to avoid long-winded phrases such as 'allelic lineage with a motif which in amino acid translation reads ...'.

The amino acids will be given in the international one-letter code and their position will be indicated in the mature protein sequence. To give an example, the main motif shared by all the *HLA–DRB1*03* alleles (where *HLA* stands for 'human leukocyte antigen, the human MHC, and the numbers following the asterisk are allelic designations) is 9-EYST-12. At the same position, the *HLA–DRB1*04* alleles have the sequence 9-EQVK-12, the *HLA–DRB1*02* alleles 9-WQPK-12, and so on. Other motifs occur at positions 26–33, 37 and 38, and 67–74, which are also the most polymorphic segments of the *DRB1* genes. The alleles within an allelic lineage differ from one another by solitary substitutions scattered throughout the rest of the sequence.

The most remarkable observation is that the motifs, although allelic lineage-specific, occur as polymorphisms in a variety of other primate species; some even occur in non-primate eutherian mammals. For example, the lineage with the EYST motif has been found not only in humans, but in common chimpanzee (*Patr*), pygmy chimpanzee (*Papa*), gorilla (*Gogo*), orangutan (*Popy*), rhesus macaque (*Mamu*), pigtail macaque (*Mane*), Japanese macaque (*Mafu*), drill (*Male*), hamadryas baboon (*Paha*), common marmoset (*Caja*), cotton-top tamarin (*Saoe*), dusky titi (*Camo*), and northern lesser bushbaby (*Gase*): hence in apes (*Patr, Papa, Gogo, Popy*), Old World monkeys (*Mamu, Mane, Mafu, Male, Paha*), New World monkeys (*Caja, Saoe, Camo*), and prosimians (*Gase*) (see O'hUigin *et al.* 1993). (The four-letter italicized symbols are abbreviations of the scientific genus and species names; for example, *Patr* stands for *Pan troglodytes*.) Among the non-primates, its occurrence has been documented in sheep (*Ovar*), anoa (Asian buffalo, *Bude*), cattle (*Bota*), banteng (*Boja*), goat (*Caae*) (see Schwaiger *et al.* 1993), and mouse (*Mumu*) (see Lundberg & McDevitt 1992).

It is, however, not just the motif that is shared by the different primate species, but the entire allelic lineage. The interspecific sharing of allelic lineages is indicated by the overall sequence similarity of corresponding genes in the different species, which is reflected in genetic distances calculated from the similarities and in dendrograms based on the genetic distances (but also on parsimony considerations). In the dendrograms, genes of a given lineage but from different species cluster together and separate from clusters (clades) formed by other allelic lineages. The MHC polymorphism, therefore, has a trans-species character (figure 1) (Klein 1987). Each species possesses between five and seven major allelic lineages, all of which are shared with at least one other species. The typing has thus far not been exhaustive enough to reveal whether some species have lost any of the major lineages, although this would not be surprising, particularly in respect of those on the endangered species list, whose populations have been drastically reduced in recent times. It is clear, however, that all the major allelic lineages have passed through many speciation phases. Several of the *DRB1* allelic lineages may be more than 40 million years (Ma) old.

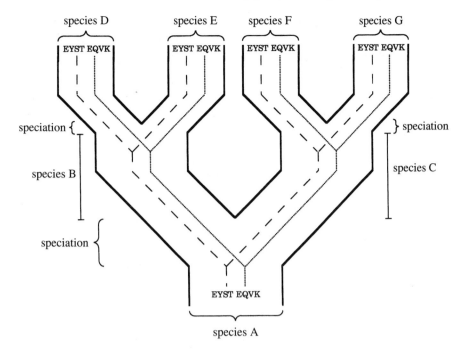

Figure 1. Passage of MHC allelic lineages through speciation phases. Three speciation events are shown. The MHC lineages (broken and dotted lines) carry characteristic motifs (amino acid residues abbreviated in the international single-letter code) that are retained through the entire evolutionary period. Not shown are the individual alleles that arise in each species from each lineage by mutations.

3. EVOLUTION BY INCORPORATION

The trans-species character of the allelic lineages is established by a special mode of the evolutionary process (Klein *et al.* 1991). Instead of the standard fixation of mutations in diverging species, the MHC allelic lineages evolve by a process that we refer to as *incorporation* (fixation within a lineage). In this process, mutations spread, one after another, through only a portion of the population and thus remain at frequencies of less than 1 over many millions of years. Hence an allele that has acquired one mutation is in a position to acquire a second before the first achieves fixation. Another group of individuals of the same species acquires a different mutation, and then a second, again without the first becoming fixed, and so on. Thus, mutation after mutation is incorporated into the lineages, which thus gradually drift apart in their sequences. It takes on average 1.3 Ma to incorporate a non-synonymous substitution into the PBR of a primate class I gene and approximately 4.5 Ma into the PBR of a class II gene. The times for the incorporation of synonymous substitutions are 3.2 Ma and 5.0 Ma, respectively (Klein *et al.* 1993). Because the average lifespan of a species is 2 Ma (Stanley 1975) and *HLA* alleles can differ by up to 55 substitutions (39 non-synonymous, 16 synonymous in *HLA–DRB1*0301* and *HLA–DRB1*0701* alleles), it is obvious that the lineages must evolve transspecifically.

It is also obvious that the time during which mutations remain as polymorphisms in a succession of populations (species) exceeds the expectations based on the coalescence theory of neutral alleles. The mean coalescence time of two neutral alleles is $2N_e$ generations, where N_e is the effective population size (Hartl & Clark 1989). The estimated N_e in the hominoid lineage in the last 20 my is 10^5 individuals (Takahata *et al.* 1994), giving an expected mean coalescence time of neutral alleles of 2×10^5 generations. Taking a generation time of 20 years for the hominoids, we obtain a mean coalescence time of two neutral alleles of 4 Ma. In reality, the mean coalescence time of MHC alleles is at least ten times higher. The obvious conclusion is that MHC genes are not neutral and that their long persistence can be explained only by invoking a form of balancing selection (Takahata 1990). The occurrence of balancing selection in the MHC genes is also indicated by the observation that the per site frequency of nonsynonymous substitutions in the PBR is several times higher than that of synonymous substitutions (Hughes & Nei 1989).

4. MHC AND PARASITES

The only known function of the classical MHC molecules is to present peptides to T lymphocytes of the immune system. It therefore seems logical to assume that the agents behind the balancing selection are the parasites, from which the immune system, with the MHC molecules at the forefront, protects the vertebrates. Involvement in the immune response is also the only postulated function of the MHC which all the vertebrates share. Neither odour-based mate (kin) selection (Potts & Wakeland 1994), nor maternal–foetal incompatibility (Gill 1994), nor any of the other selection pressures proposed for the maintenance of MHC polymorphism have the ubiquitous distribution among the vertebrate taxa that the parasite-targeted immune response does.

Assuming that the generation, not only of allelic lineages with their lineage-specific motifs, but also of MHC polymorphism in general, is indeed driven and maintained by parasites, several difficult questions arise. Foremost among them is why the MHC polymorphism evolves in a trans-species manner. Undoubtedly, the answer to this question will emerge from the study of host–parasite interactions. We divide parasites in this regard into three categories, limiting our discussion to human parasites only. In the first category are parasites that have become associated with humans only in the past 10 000 years (Cameron 1956; Johnson 1986; Nelson 1988). They are often highly virulent and have been responsible for past and present epidemics and pandemics. Because they have been the cause of high mortality in human populations, they are widely believed to have exerted strong selection pressure on the evolution of MHC polymorphism. Examples include *Plasmodium falciparum*, *Trypanosoma gambiense*, *Leishmania donovani*, *Mycobacterium tuberculosis*, *Yersinia pestis*, smallpox virus, rubella virus, and influenza virus. All these organisms are believed to have become human parasites after the advent of agriculture and domestication of animals, which led to an increasing population density and aggregation of humans in large communities. For several of them, there is indeed good experimental evidence for their recent association with *Homo sapiens*. For example, sequencing of the small-subunit ribosomal RNA gene from a variety of *Plasmodium* species has revealed that *P. falciparum*, the causative agent of falciparan malaria, is so closely related to avian *Plasmodium* species that a lateral transfer of the parasite from birds to humans as recently as the start of the agricultural revolution is the most plausible explanation of its origin (Waters *et al.* 1993). In a few cases, an explanation for increased virulence of the parasite at the time of the transfer is also available. Thus, in *Yersinia pestis* (the causative agent of the bubonic plague) two mutations, one at a chromosomal locus and the other in a plasmid, may have produced hypervirulent strains of these bacteria which, when transferred from rats and fleas to humans, caused the plague epidemics (Rosqvist *et al.* 1988).

5. PARASITES AND MHC POLYMORPHISM

We argue that these recent parasites have had little effect on MHC polymorphism. We base our argument on two observations. The first is that the great majority of non-synonymous PBR substitutions differentiating *HLA* alleles and MHC lineages are much older than 10 000 years and therefore could not have been incorporated in response to this category of parasites. If a non-synonymous PBR substitution is incorporated into the human population once every 1.3–5 Ma, obviously most alleles that differ by a single substitution, and almost certainly all the alleles that differ by two or more substitutions and were not generated by complex mutations or by recombination,

must have been established long before the beginning of the Neolithic revolution. Their persistence in the hominoids therefore could not have been caused by parasites that invaded the human population less than 10 000 years ago. The second observation is that the high virulence of these parasites often paralyses the host's immune system, which then provides little protection against the invaders. Often the parasite kills the host quickly, before the latter can mobilize its anticipatory immune system. There is thus very little selection on any component of the immune system, including the MHC genes.

Our exclusion of the first category of parasites from our search for the driving force of MHC polymorphism may seem to be contradicted by the studies on falciparan malaria. Several such studies have been carried out, but the largest and the most publicized of them was by Hill and his colleagues (1991) on Gambian black populations. The main conclusions of this study were that children under ten years of age and carrying the *HLA-B53* allele have a 40% lower chance of succumbing to either severe malarial anaemia or cerebral malaria than those lacking this allele. Similarly, children carrying the *DRB1*1302 – DQB1*0501* haplotype have a 50% lower chance of dying from malarial anaemia than children with other haplotypes. The *DRB1*1302 – DQB1*0501* haplotype, however, offers no protection against cerebral malaria and neither it, nor the *HLA-B53* allele, provides any protection from the initial infection by *P. falciparum*. The study, even if confirmed by extension to other populations, cannot be regarded as evidence that *P. falciparum* is one of the agents driving MHC polymorphism, nor that it is responsible for the divergence of allelic lineages. Hill *et al.* (1991) suggest that, during the past 5000 years, malaria has driven the frequencies of *HLA-B53* among African blacks from the base level of 1% up to 22% in some populations (Allsopp *et al.* 1992). If so, the increase would have been accompanied by a corresponding decrease in frequencies of other *HLA* alleles and hence by an overall loss rather than gain of MHC polymorphism, especially as *HLA-B53* heterozygotes are no more resistant to *P. falciparum* than *HLA-B53* homozygotes. In fact, it has been suggested (Carter *et al.* 1992) that the effect Hill and his colleagues observed was a manifestation of decreased rather than enhanced immune response triggered by *P. falciparum* infection in the presence of *HLA-B53* compared with response in the presence of other *HLA-B* alleles, and that the heightened reactivity resulted in autoaggression.

Into the second category we place parasites whose evolution appears to follow Fahrenholz's rule that common ancestors of present-day parasites were themselves parasites of the common ancestors of present-day hosts (Fahrenholz 1913). Because they have speciated in parallel with their hosts, the immune system, including the MHC molecules, must have coevolved with them. The parasites could therefore have been the source of selection pressure for periods that can be measured in millions of years. An example of this category are the papova viruses. DNA sequence comparisons of papova viruses BKV,

SV40 and polyoma virus, derived from humans, monkeys and mice, respectively, yield a phylogenetic tree matching that of their hosts (Soeda *et al.* 1980; Shadan & Villarreal 1993) and suggest a host–parasite coevolution for over 30 Ma. Long-term coevolution of human ancestors with their parasites is also indicated for some trematode worms of the genus *Schistosoma* (Despres *et al.* 1992) and for non-falciparan *Plasmodium* species (Waters *et al.* 1993). Numerous cases of parasite–host coevolution have been compiled by Brooks & McLennan (1993).

Into the third category we assign organisms so well adapted that they normally not only cause no measurable damage, but their presence is often beneficial to their hosts. Only when the host becomes immunologically compromised, for example in individuals infected with the human immuno-deficiency virus or individuals immunodeficient because of a genetic defect, is the true nature of these organisms revealed. Then the harmless com-mensals turn into *bona fide* parasites and cause an opportunistic infection resulting more often than not in death This behaviour indicates that, in a healthy individual, these opportunistic parasites are kept at bay by the immune system, which includes the MHC molecules. It has been estimated that a healthy human houses about 10^{12} bacteria on the skin, 10^{10} in the mouth, and 10^{14} in the digestive tract (Mims 1977). It must be a tremendous challenge to the immune system to survey this mass of *Gastarbeitern* and make sure that their behaviour does not violate the body's norms. The human opportunistic parasites include viruses (cytomegalovirus, herpes simplex virus, varicella–zoster virus), bacteria (*Escherichia coli*, *Pseudomonas*, *Streptococcus*), fungi (*Candida*) and protozoa (*Pneumocystis carini*). Many of these organ-isms are widespread in nature and are of an old age. *E. coli*, for example, is a normal constituent of the gut flora in a wide range of warm-blooded animals. It is estimated to have diverged from *Salmonella*, its closest relative, between 120 and 160 Ma ago (Ochman & Wilson 1987); during all that time, it has presumably remained a commensal and an occasional parasite of a great many different host species. The population of *E. coli* has been shown to be clonal (Selander & Levin 1980; Ochman & Selander 1984); the clones, as in other parasites, are believed to be of ancient origin. (Some of the *Trypanosoma cruzi* clones, for example, are estimated to have been isolated from each other for 40–50 Ma BP; see Tibayrenc *et al.* (1986).) It is hard to imagine that such a ubiquitous mass of parasites that has presented so great a challenge to their hosts for so long a period should not become a major driving force in the evolution of the vertebrate immune system and in particular of the MHC.

We propose, therefore, that parasites of the second and third categories are primarily responsible for the evolution and for the properties of MHC polymorph-ism. They are in a position to provide steadfast pressure on the MHC genes to diversify and for the variants to endure for periods exceeding those normally allotted to a species. Parasites in the first category, on the other hand, have little, if any,

influence on MHC polymorphism and may actually be reducing it. If our reasoning is correct, two further deductions can be made. First, one of the explanations for the scantiness of empirical evidence regarding the effect of parasites on MHC polymorphism is that the wrong parasites (those of the first category) have been studied. Second, it will be necessary to design experiments in which the selection pressure exerted on the MHC will be measured by taking commensals or adapted parasites from a host of one MHC haplotype and transferring them to a slightly immunocompromised host of another haplotype.

6. PARASITES AND ALLELIC LINEAGES

If the reason for the trans-species evolution of MHC polymorphism is the long-lasting host–parasite asso-ciation and coevolution of the two protagonists, how then does the existence of the allelic lineages fit into this concept? Could the lineages not simply be the result of the stochasticity of allelic genealogy? We believe not, for the following reason. According to the theory of gene genealogy, all the genes of a given generation coalesce into a single, most recent common ancestor $4N_e$ generations ago (Hartl & Clark 1989). When the coalescence process is dissected into separate steps, it turns out that the coalescence of all the genes to two ancestors takes $2N_e$ generations and the coalescence of these two into a single ancestor another $2N_e$ generations. In the *HLA–DRB* system, an initial single locus has duplicated repeatedly and produced nine paralogous loci (*DRB1* to *DRB9*). Most of the duplications can be dated approximately to the time of the primate emergence on the evolutionary scene; only one duplication may have occurred somewhat later (Klein *et al.* 1991). Had the emergence of allelic lineages predated the initial duplication, genes at different loci should be more closely related to some of the lineages, but this is not the case. From the genetic distances between the lineages, we estimate that all the lineages began diverging at about the same time. This observation contradicts the expectation under a purely stochastic process, according to which the divergence of the first two lineages should have taken half the time of the total divergence of all the lineages.

We believe, therefore, that there is more to the existence of MHC allelic lineages than genealogical stochasticity. The differentiation into lineages may have been assisted by one of two processes. First, the discontinuity in the spectrum of MHC variability manifested in the existence of the lineages may be a reflection of discontinuities in the parasite spectrum manifested in the existence of different species and clones within a 'species' (Klein 1991). Second, the lineages may be manifestations of a strategy to keep parasites from spreading through the entire popula-tion. A parasite that has learned to breach defences built on antigen presentation through one lineage of MHC molecules will not be able to avoid the immune response in individuals of the same species carrying other MHC lineages, if the lineages differ from one another to a great extent. We prefer the former argument because the latter smacks of group selection.

The existence and perpetuation of motifs characterizing individual lineages may be explained in the same manner. Sharing of motifs between MHC genes and proteins may, however, occur in two ways: by descent and by convergence. There are good reasons to believe that shared motifs characterizing the *DRB1* lineages in different catarrhine primates owe their similarity to the common origin of the genes within the lineage. The strongest evidence for the common origin hypothesis is the sharing by these genes of the same inserts of *Alu* and retroviral elements (Y. Satta & J. Klein, manuscript in preparation). The occasional occurrence of a *DRB1* motif in non-primate mammals (such as the presence of the EYST motif in artiodactyls and rodents), on the other hand, must be the result of convergence. This conclusion is supported by two observations. First, we have examined some 500 *DRB* alleles and allelic pairs differing by single, double, and treble replacements. The location and type of replacements differentiating each member of the pair were scored and the data were used to measure (i) the relative frequency of replacement at each of the PBR sites and (ii) the types of replacement that occur. Some PBR sites were observed to undergo replacement at a much higher frequency than others; at many PBR sites, replacements are limited to a few amino acids. When we applied these observations to a coalescence model to simulate lineage divergences, we found that highly diverged alleles may resemble each other at a number of motif-bearing sites and hence that convergence has occurred (O'hUigin 1994).

Based on the high degree of shared codon usage in the DNA encoding the motifs, Lundberg & McDevitt (1992) have argued that polymorphisms in the motif-encoding DNA segments have directly descended to mouse and humans from the eutherian ancestor, whereas the lineages in each species have been generated *de novo* through recombination. However, because the third-position GC content in the class II genes examined by Lundberg & McDevitt is quite high (70–80%) and because GC content is the major determinant of mammalian codon usage (Aota & Ikemura 1986), the shared codon usage may be attributable to the high GC content.

Second, as mentioned above, the *DRB1* lineages apparently arose after the duplication of the ancestral *DRB* gene in the primates. They could not, therefore, have existed in the common ancestor of primates, artiodactyls, and rodents. The same conclusion applies to the motifs diagnostic of the individual lineages; these must have arisen independently in the three mammalian orders.

7. MHC AND SELF

We have limited our discussion of MHC polymorphism to the stimuli provided by the non-self world, the parasites. Of course, the MHC molecules interact most of the time not with non-self, but with self peptides; one may wonder, therefore, whether some minor, or even major, influence on the polymorphism may not come from the host itself. Space does not allow us to go into any details, except to state our position on this topic. A failure to bind a critical non-self peptide may endanger the life of an individual; a failure to bind a self peptide is probably to have no consequences at all because the worst that can happen is that the molecule will not be expressed on the cell surface. However, this is unlikely, because there will always be other self peptides that will bind, and even if a molecule were not expressed, the encoding element would eventually turn into a pseudogene and be replaced by another gene. In either case, there would not be much influence on the polymorphism of the functional MHC genes. It is true, however, that little is still known about the process of T-cell selection by MHC molecules in the thymus and it is therefore possible that this view will need to be modified once the actual mechanism of the selection process is elucidated.

We thank Ms Lynne Yakes for editorial assistance and Dr Robert Slade for critical reading of the manuscript.

REFERENCES

Allsopp, C.E.M., Harding, R.M., Taylor, C., Bunce, M., Kwiatkowski, D., Anstey, N., Brewster, D., McMichael, A.J., Greenwood, B.M. & Hill, A.V.S. 1992 Interethnic genetic differentiation in Africa: HLA class I antigens in the Gambia. *Am. J. hum. Genet.* **50**, 411–421.

Aota, S. & Ikemura, T. 1986 Diversity in G + C content at the third position of codons in vertebrate genes and its cause. *Nucl. Acids Res.* **14**, 6345–6355.

Brooks, D.R. & McLennan, D.A. 1993 *Parascript, Parasites and the language of evolution.* Washington: Smithsonian Institution Press.

Cameron, T.W.M. 1956 *Parasites and parasitism.* London: Methuen & Co.

Carter, R., Schofield, L. & Mendis, K. 1992 HLA effects in malaria: Increased parasite-killing immunity or reduced immunopathology? *Parasitol. Today* **8**, 41–42.

Despres, L., Imbert-Establet, D., Combes, C. & Bonhomme, F. 1992 Molecular evidence linking hominoid evolution to recent radiation of schistosomes (Platyhelminthes: Trematoda). *Molec. phylogenet. Evol.* **1**, 295–304.

Fahrenholz, H. 1913 Ectoparasiten und Abstammungslehre. *Zool. Anz., Leipzig* **41**, 371–374.

Germain, R.N. 1994 MHC-dependent antigen processing and peptide presentation: Providing ligands for T lymphocyte activation. *Cell* **76**, 287–299.

Gill, T.J. III 1994 Reproductive immunology and immunogenetics. In *The Physiology of reproduction* (ed. E. Knobil & J. D. Neil), 2nd edn, pp. 783–812. New York: Raven Press.

Hartl, D.L. & Clark, A.G. 1989 *Principles of population genetics,* 2nd edn. Sunderland, Massachusetts: Sinauer.

Hill, A.V.S., Allsopp, C.E.M., Kwiatkowski, D., Anstey, N.M., Twumasi, P., Rowe, P.A., Bennett, S., Brewster, D., McMichael, A.J. & Greenwood, B.M. 1991 Common West African HLA antigens are associated with protection from severe malaria.; *Nature, Lond.* **352**, 595–600.

Hughes, A.L. & Nei, M. 1989 Nucleotide substitution at major histocompatibility complex class II loci: Evidence for overdominant selection. *Proc. natn. Acad. Sci. U.S.A.* **86**, 958–962.

Johnson, R.B. 1986 Human disease and the evolution of pathogen virulence. *J. theor. Biol.* **122**, 19–24.

Klein, J. 1986 *Natural history of the major histocompatibility complex.* New York: John Wiley.

Klein, J. 1987 Origin of major histocompatibility complex polymorphism: The trans-species hypothesis. *Hum. Immun.* **19**, 155–162.

Klein, J. 1989 Are invertebrates capable of anticipatory immune response? *Scand. J. Immun.* **29**, 499–505.

Klein, J. 1991 Of HLA, tryps, and selection: An essay on coevolution of MHC and parasites. *Hum. Immun.* **30**, 247–258.

Klein, J., Satta, Y., O'hUigin, C., Mayer, W.E. & Takahata, N. 1991 Evolution of the primate *DRB* region. In *HLA 1991* (ed. T. Sasazuki), vol. 2, pp. 45–56. (Proceedings of the 11th International Histocompatibility Workshop, Yokohama, Japan.) Oxford University Press.

Klein, J., Satta, Y., O'hUigin, C. & Takahata, N. 1993 The molecular descent of the major histocompatibility complex. *A. Rev. Immun.* **11**, 269–295.

Lundberg, A.S. & McDevitt, H.O. 1992 Evolution of major histocompatibility complex class II allelic diversity: Direct descent in mice and humans. *Proc. natn. Acad. Sci. U.S.A.* **89**, 6545–6549.

Marsh, S.G.E. & Bodmer, J.G. 1993 *HLA* class II nucleotide sequences, 1992. *Immunogenetics* **37**, 79–94.

Martin, R.D. 1993 Primate origins: plugging the gaps. *Nature, Lond.* **363**, 223–234.

Miller, J.F.A.P. 1992 The key role of the thymus in the body's defense strategies. *Phil. Trans. R. Soc. Lond.* B **337**, 105–124.

Mims, C.A. 1977 *The pathogenesis of infectious disease.* London: Academic Press.

Neefjes, J.J. & Momburg, F. 1993 Cell biology of antigen presentation. *Curr. Opin. Immun.* **5**, 27–34.

Nelson, G.S. 1988 Parasitic zoonoses. In *The biology of parasitism*, pp. 13–41. New York: Alan R. Liss.

Ochman, H. & Wilson, A.C. 1987 Evolution in bacteria: Evidence for a universal substitution rate in cellular genomes. *J. molec. Evol.* **26**, 74–86.

Ochman, H. & Selander, R.K. 1984 Evidence for clonal population structure in *Escherichia coli. Proc. natn. Acad. Sci. U.S.A.* **81**, 198–201.

O'hUigin, C. 1994 Quantifying the degree of convergence in primate Mhc-DRB genes. *Immunol Rev.* (In the press.)

O'hUigin, C., Bontrop, R. & Klein, J. 1993 Nonhuman primate MHC-*DRB* sequences: a compilation. *Immunogenetics* **38**, 165–183.

Potts, W.K. & Wakeland, E.K. 1994 Evolution of MHC genetic diversity: A tale of incest, pestilence and sexual preference. *Trends Genet.* **9**, 408–413.

Rammensee, H.G., Falk, K. & Rötzschke, O. 1993 MHC molecules as peptide receptors. *Curr. Opin. Immun.* **5**, 35–44.

Rosqvist, R., Skurnik, M. & Wolf-Watz, H. 1988 Increased virulence of *Yersinia pseudotuberculosis* by two independent mutations. *Nature, Lond.* **334**, 522–525.

Schwaiger, F.-W., Weyers, E., Epplen, C., Brün, J., Ruff, G., Crawford, A. & Epplen, J.T. 1993 The paradox of MHC-*DRB* exon/intron evolution: α-helix and β-sheet encoding regions diverge while hypervariable intronic simple repeats coevolve with β-sheet codons. *J. molec. Evol.* **37**, 260–272.

Selander, R.K. & Levin, B.R. 1980 Genetic diversity and structure in *Escherichia coli* populations. *Science, Wash.* **210**, 545–547.

Shadan, F.F. & Villarreal, L.P. 1993 Coevolution of persistently infecting small DNA viruses and their hosts linked to host-interactive regulatory domains. *Proc. natn. Acad. Sci. U.S.A.* **90**, 4117–4121.

Soeda, E., Maruyama, T., Arrand, J.R. & Griffin, B.E. 1980 Host-dependent evolution of three papova viruses. *Nature, Lond.* **285**, 165–167.

Stanley, S.M. 1975 A theory of evolution above the species level. *Proc. natn. Acad. Sci. U.S.A.* **72**, 646–650.

Takahata, N. 1990 A simple genealogical structure of strongly balanced allelic lines and trans-species evolution of polymorphism. *Proc. natn. Acad. Sci. U.S.A.* **87**, 2419–2423.

Takahata, N., Satta, Y. & Klein, J. 1994 Divergence time and population size in the lineage leading to modern humans. *Theor. Popul. Biol.* (In the press.)

Tibayrenc, M., Ward, P., Moya, A. & Ayala, F.J. 1986 Natural populations of *Trypanosoma cruzi*, the agent of Chagas disease, have a complex multiclonal structure. *Proc. natn. Acad. Sci. U.S.A.* **83**, 115–119.

Waters, A.P., Higgins, D.G. & McCutchan, T.F. 1993 The phylogeny of malaria: A useful study. *Parasitol. Today* **9**, 246–252.

Discussion

C. A. MIMS (*Sheriff House, Ardingly, West Sussex, U.K.*). Professor Klein suggests that in the case of infections that appeared in humans more than 10 000 years ago, the MHC has not had time to respond by developing appropriate molecules for the presentation of the relevant microbial antigens. My question is about time. We know that host genetic resistance can develop quite rapidly under pressure from a virulent infection. Host and virus changes have been thoroughly charted in myxomatosis in the Australian rabbit, an infectious disease that evolved in a highly susceptible host. Genetically resistant rabbits appeared within 10–15 years, which, after allowing for differences in generation times between humans and rabbits, is far less than the 10 000 years you suggested for human infections. Also, although here the picture is complicated by the influence of nutrition, poverty, housing, it seems probable that in Europeans during the past few hundred years there has been a significant selection for genotypes resistant to tuberculosis. Of course, it is not known to what extent any of these host genetic changes are based on MHC, but it leads me to ask why the MHC should take so long in responding to life-threatening infections.

J. KLEIN. I wanted to say that the MHC polymorphism now present in the global population is more than 10 000 years old and must have therefore been established by selection pressures other than those that appeared after the neolithic revolution. It was not my intention to deny that new variants keep emerging in local populations and that their fate may be influenced by positive selection exerted by parasites. But the time needed for their incorporation as polymorphisms into the global population is, on average, longer than 0.5 million years. The other point I was trying to make was that the MHC (or, more precisely, the anticipatory immune system) may not be able to protect us (or the rabbits) from highly virulent parasites. Genes other than the MHC may decide who is to be resistant and who susceptible, and the infection stops when the parasites do not find any more susceptible individuals to infect.

J. LINES (*London School of Hygiene and Tropical Medicine, U.K.*). Professor Klein raised the question of why we do not see more obvious associations between the MHC and infectious disease in natural populations. However, suppose the relationship between host and parasite were a matching allele system, and that the allele frequencies of both partners were at selective equilibrium. At equilibrium, genotype fitnesses are by definition equal, and so at the population level there would be no visible association between a given host genotype and disease in general. Particular parasite genotypes would be associated with particular host genotypes, but this would only be seen by scoring the genotype of both parasite and host in individual infections. Have any such surveys, taking account of the genotype of both partners, been carried out with the infections that Professor Klein showed us in his table?

J. KLEIN. In all the studies that I am aware of, the parasite is regarded as being homogeneous and only the frequencies of MHC alleles of the susceptible and resistant individuals are compared.

J. DEUTSCH (*Zoology Department, University of Cambridge, U.K.*). If one were able to characterize the MHC allelic lineages in two closely related species, would it be possible to estimate the population size of initial populations at speciation based upon the proposition of allelic lineages shared by the species?

J. KLEIN. One of the main reasons why we have been sequencing MHC alleles of related primate species is to estimate the sizes of the founding populations using the

population genetics theory of allelic genealogy and models of computer simulations. The results obtained thus far indicate that for *Homo sapiens*, the founding effective population size was between 10 000 and 100 000 individuals (see Klein *et al. Trends Genet.* **6**, 7–11 (1990); Klein *et al. J. med. Primatol.* **22**, 57–64 (1993); Klein *et al. Scient. Amer.* December 1993, pp. 56–62.)

A. L. HUGHES (*Department of Biology, Pennsylvania State University, U.S.A.*). I think it need not be true that *Plasmodium Falciparum* has become a human parasite only recently.

J. KLEIN. My inclusion of *Plasmodium falciparum* in the group of recent parasites is based on the data of A. P. Waters and colleagues (*Proc. natn. Acad. Sci. U.S.A.* **88**, 3140–3144 (1991)). On the basis of sequence analysis of small-subunit ribosomal RNA genes, these investigators reported that *P. falciparum* forms a monophyletic group with avian *Plasmodium* species, highly divergent from primate and rodent *Plasmodium* parasites. They argue that *P. falciparum* and avian *Plasmodium* species shared a common ancestor relatively recently and hence that their results are consistent with the hypothesis of a recent acquisition of *P. plasmodium* by humans, possibly at the time a switch-over to an agricultural-based lifestyle occurred (see also Livingstone *Am. Anthropol.* **60**, 533–560 (1958)).

10

Natural selection at the class II major histocompatibility complex loci of mammals

AUSTIN L. HUGHES, MARIANNE K. HUGHES, CARINA Y. HOWELL AND
MASATOSHI NEI

*Department of Biology and Institute of Molecular Evolutionary Genetics, The Pennsylvania State University, University Park,
PA 16802, U.S.A.*

SUMMARY

The role of natural selection at major histocompatibility complex (MHC) loci was studied by analysis of molecular sequence data from mammalian class II MHC loci. As found previously for the class I MHC molecule and a hypothetical model of the class II molecule, the rate of non-synonymous nucleotide substitution exceeded that of synonymous substitution in the codons encoding the antigen recognition site of polymorphic class II molecules. This pattern is evidence that the polymorphism at these loci is maintained by a form of balancing selection, such as overdominant selection. By contrast, in the case of monomorphic class II loci, no such enhancement of the rate of non-synonymous substitution was observed. Phylogenetic analysis indicates that, in contrast to monomorphic ('non-classical') class I MHC loci, some monomorphic class II loci of mammals are quite ancient. The *DMA* and *DMB* loci, for example, diverged before all other known mammalian class II loci, possibly before the divergence of tetrapods from bony fishes. Analysis of the patterns of sharing of polymorphic residues at class II MHC loci by mammals of different species revealed that extensive convergent evolution has occurred at these loci; but no support was found for the hypothesis that MHC polymorphisms have been maintained since before the divergence of orders of eutherian mammals.

1. INTRODUCTION

The major histocompatibility complex (MHC) is a multi-gene family whose products are cell-surface glycoproteins that function to present intracellularly processed peptides to T cells. The MHC genes are remarkable for several reasons. Their most striking characteristic is the high level of polymorphism found at certain MHC loci, which are among the most polymorphic known in organisms. Furthermore, certain of these loci are characterized by 'trans-species' polymorphism (Lawlor *et al.* 1988; Mayer *et al.* 1988); that is, polymorphic allelic lineages may be shared by related species, such as human and chimpanzee, having apparently been maintained in both species since before speciation. By contrast, certain MHC loci are monomorphic or nearly so, and the function of most of the monomorphic MHC loci is unknown.

In the following, after a brief introduction to the structure and function of MHC molecules, we present evidence from statistical analysis of DNA sequence data that addresses a number of questions regarding MHC evolution. While reviewing previous studies of class I MHC molecules, we present the results of new analyses of class II MHC sequence data. We address the following questions: (1) the mechanism of

maintenance of MHC polymorphism; (2) the evolution of monomorphic class II MHC loci; and (3) the contributions of common ancestry and convergent evolution to sharing of polymorphic amino acid motifs among different mammalian species. Because extensive sequence data are available from orthologous class II loci from several orders of mammals, data from the class II MHC are particularly appropriate for addressing this last question.

2. STRUCTURE AND FUNCTION OF MHC MOLECULES

Class I MHC molecules, which are expressed on almost all kinds of nucleated cells, present peptides to cytotoxic T cells (Bjorkman & Parham 1990; Klein 1986). Class II molecules have a much more restricted expression, being expressed primarily on antigen-presenting cells of the immune system. The complex of self class II MHC molecule and foreign peptide on an antigen-presenting cell is recognized by a helper T cell; the latter releases lymphokines that stimulate B cells (which produce antibodies) and macrophages to respond to the infection.

Both class I and class II molecules are heterodimers with four extracellular domains, but they achieve a similar molecular structure in different ways. The class I α chain includes three extracellular domains

$(\alpha_1, \alpha_2, \alpha_3)$. The α_3 domain associates non-covalently with β_2-microglobulin. β_2-microglobulin is encoded by a gene outside the MHC region; but the gene has a distant evolutionary relation to MHC genes. Portions of the α_1 and α_2 domains form a groove at the top of the molecule called the antigen recognition site (ARS), in which peptides are bound and presented to T cells (Bjorkman *et al.* 1987*a*,*b*). The ARS groove is formed by two α helices bounding a β pleated sheet.

The class II heterodimer consists of an α chain with two extracellular domains (α_1 and α_2) and a β chain with two extracellular domains (β_1 and β_2), both encoded by genes within the MHC region. A hypothetical structure for the class II molecule was initially proposed on the basis of analogy with the class I structure (Brown *et al.* 1988). Recently, the structure of the HLA–DR1 class II heterodimer has been worked out (Brown *et al.* 1993), and it is similar to the previous hypothetical model. The class II molecule has an ARS similar to that of the class I molecule; but in this case one α helix and about half of the β pleated sheet are contributed by the α_1 domain of the α chain, while the β_1 domain of the β chain contributes the other α helix and the remainder of the β pleated sheet.

The class II MHC genes of eutherian mammals are arranged in a number of separate chromosomal regions, each of which typically contains at least one α chain gene and one or more β chain genes. The genes encoding α chains are generally designated *A* (e.g. *DRA*), and those encoding β chains are designated *B* (e.g. *DRB1*). In humans, there are three regions including polymorphic class II loci, designated *DR*, *DQ* and *DP*. Each of these regions contains at least one polymorphic β chain locus. In the *DR* region, there are four functional β chain loci (*DRB1*, *DRB3*, *DRB4* and *DRB5*). Between one and three of these β chain loci are present in an individual haplotype. The *DQ* and *DP* regions each contain a single functional β chain locus (*DQB1* and *DPB1*, respectively). Only in the *DQ* region is there a substantial degree of polymorphism in the α chain gene (*DQA1*). However, sequences of a number of alleles are also available for *DPA1*.

In a number of other eutherian mammals, homologues of the human class II loci have been discovered, and these are usually named following their human homologues. However, because the class II loci of the mouse were discovered independently of those of humans, a different system of nomenclature has been used for genes in the mouse MHC (called the *H-2* complex). The symbols *a* and *b* are used respectively for class II α and β chain genes. The regions corresponding to human *DR* and *DQ* are known as *E* and *A*, respectively; the mouse lacks functional *DP* homologues. Thus the mouse locus homologous to *HLA–DRA* is called *H-2Ea*.

3. MAINTENANCE OF MHC POLYMORPHISM

When the function of the class I MHC in immune recognition was worked out, its role in antigen presentation suggested to Doherty & Zinkernagel

(1975) a mechanism whereby overdominant selection (heterozygote advantage) might operate at MHC loci. They had evidence that products of different alleles at MHC loci differ in their ability to bind and present specific foreign peptides. Given such differences, in a population exposed to a variety of pathogens, an individual heterozygous at all or most MHC loci would presumably be able to present more kinds of foreign peptides than a homozygote. Thus, a heterozygote would be resistant to a wider array of pathogens than would a homozygote.

Hughes & Nei (1988, 1989*a*) approached the question of selection at MHC loci by comparing rates of synonymous and non-synonymous (amino acid altering) nucleotide substitution. They used the method of Nei & Gojobori (1986) to estimate the number of synonymous nucleotide substitutions per synonymous site (d_S) and the number of non-synonymous nucleotide substitutions per non-synonymous site (d_N) in different regions of MHC genes. Hughes & Nei (1988) reasoned that, if MHC polymorphisms are maintained by overdominant selection relating to peptide binding and thus to pathogen resistance, such selection would act to favour amino acid differences in the ARS. Therefore d_N should exceed d_S in the codons encoding the ARS. This pattern is the reverse of that seen in most functional genes, in which d_S exceeds d_N. This occurs because most amino acid changes disrupt protein structure and are harmful to the organism, and thus most non-synonymous mutations will be eliminated by conservative or 'purifying' natural selection.

As predicted, Hughes & Nei (1988) observed values of d_N significantly higher than d_S in the class I ARS. By contrast, they found that d_S generally exceeded d_N in the rest of the class I gene. Thus, the class I gene is subject to overdominant selection or some similar form of balancing selection, favouring diversity in the ARS, whereas the remainder of the molecule is subject to purifying selection. Similar results were found for class II genes (Hughes & Nei 1989*b*). In the latter, although a crystallographic structure was not available, a hypothetical structure based on analogy with the class I molecule had been proposed (Brown *et al.* 1988).

Here we present the results of new analyses of class II MHC sequences from human and mouse, based on the HLA–DR1 structure (Brown *et al.* 1993). Tables 1 and 2 show estimates for mean d_S and d_N in pairwise comparisons among alleles from polymorphic human and mouse class II MHC. These values were estimated separately for the ARS codons; for the remainder of D1 (α_1 or β_1), excluding the ARS codons; and for D2 (α_2 or β_2). At polymorphic β chain loci in both species, d_N is enhanced in the ARS (table 1). The same pattern is seen at the *Aa* locus of the mouse, but not at *DPA1* or *DQA1* of humans (table 1). For *DPA1* there is no significant difference between d_S and d_N in the ARS (table 1). Overall the level of sequence divergence among *DPA1* alleles is low compared with that found at other human polymorphic class II loci. For example, the d_S values for the non-ARS portion of Domain 1 and for Domain 2 are lower for

Table 1. *Mean numbers of synonymous (d_S) and non-synonymous (d_N) nucleotide substitutions per 100 sites among class II MHC alleles of human and mouse*

locus	(n)	antigen recognition site		remainder, D1		D2	
		d_S	d_N	d_S	d_N	d_S	d_N
β chain loci							
mouse							
Ab	(9)	0.6 ± 0.8	21.7 ± 4.0***	4.3 ± 1.9	6.0 ± 1.3	9.2 ± 2.5	1.2 ± 0.5**
Eb	(6)	5.2 ± 3.8	22.9 ± 4.4**	1.3 ± 1.0	3.1 ± 0.9	1.0 ± 0.8	0.5 ± 0.3
human							
DPB1	(6)	1.7 ± 2.4	9.8 ± 3.0*	1.8 ± 1.4	1.9 ± 0.7	4.3 ± 1.9	0.7 ± 0.4
DQB1	(9)	7.3 ± 4.2	19.3 ± 3.6*	11.1 ± 2.9	5.7 ± 1.2	5.4 ± 2.0	1.7 ± 0.6
DRB1	(23)	6.1 ± 3.9	24.7 ± 3.5***	9.5 ± 2.5	3.4 ± 0.8*	6.1 ± 1.7	2.3 ± 0.6*
DRB3	(4)	4.5 ± 4.5	10.3 ± 3.3	1.3 ± 1.3	1.2 ± 0.7	4.3 ± 1.9	0.9 ± 0.5
DRB5	(4)	4.3 ± 4.4	10.8 ± 3.7	2.7 ± 1.9	0.8 ± 0.6	2.7 ± 1.6	0.6 ± 0.4
all DRB	(32)	5.6 ± 3.9	26.9 ± 3.3***	9.7 ± 2.2	3.8 ± 0.7*	7.4 ± 1.7	2.7 ± 0.6**
α chain loci							
mouse							
Aa	(6)	3.6 ± 3.3	15.8 ± 3.7*	2.7 ± 1.6	4.1 ± 1.0	7.6 ± 2.2	0.7 ± 0.4***
human							
DPA1	(4)	0.0 ± 0.0	2.1 ± 1.5	5.1 ± 2.6	0.7 ± 0.5	2.4 ± 1.4	0.7 ± 0.4
DQA	(6)	23.2 ± 12.0	7.2 ± 2.7	9.7 ± 3.0	8.4 ± 1.7	4.1 ± 1.7	1.4 ± 0.5

Tests of the hypothesis that $d_S = d_N$: *$P < 0.05$; **$P < 0.01$; ***$P < 0.001$. D1 is β_1 or α_1 domain; D2 is β_2 or α_2 domain. Comparison of all *DRB* includes one *DRB4* sequence. n, Number of sequences.

comparisons among *DPA1* alleles than for most comparisons among alleles at other human class II loci analysed (table 1). Thus, the *DPA1* alleles seem to have diverged from each other more recently than have alleles at the other loci. These results suggest that *DPA1* polymorphism is probably selectively neutral.

The case of *DQA1* is somewhat different. At this locus, d_S and d_N do not differ significantly in the ARS (table 1). However, as measured by d_S values among alleles, *DQA1* alleles are quite divergent from each other (table 1), suggesting that polymorphism at this locus has been maintained for a long time. The existence of trans-species polymorphism at this locus in humans and other Old World primates was

supported by phylogenetic analyses (Gyllensten & Erlich 1989; Nei & Rzhetsky 1991). At this locus in humans and great apes, alleles belong to four lineages, within which there are minor variants; and these lineages seem to have been maintained at least since before the divergence of human and gorilla (Nei & Rzhetsky 1991). Thus at the *DQA1* locus of higher primates a somewhat different type of balancing selection may be occurring than at other polymorphic MHC loci. This selection may act to maintain the four allelic lineages but may not favour major new allelic forms of the ARS. By contrast, the homologous *Aa* locus of the mouse shows a pattern of nucleotide substitution similar to that found at polymorphic β

Table 2. *Numbers of synonymous (d_S) and non-synonymous (d_N) nucleotide substitutions per 100 sites in comparisons between orthologous class II MHC loci of human and mouse*

comparisons	(n)	antigen recognition site		remainder, D1		D2	
		d_S	d_N	d_S	d_N	d_S	d_N
β chain loci							
monomorphic loci							
DMB vs. Mb	(2)	56.8 ± 25.3	31.0 ± 9.1[b]	76.9 ± 20.6	17.2 ± 3.6**	61.8 ± 13.3	11.9 ± 2.5***
DOB vs. Ob	(2)	45.2 ± 20.4	7.8 ± 3.9	58.1 ± 15.7	13.7 ± 3.1**	84.6 ± 18.7	22.7 ± 3.6**
polymorphic loci							
DQB vs. Ab	(18)	54.1 ± 21.7	32.5 ± 4.9[c]	30.4 ± 7.9	17.6 ± 8.9	53.6 ± 11.0	10.5 ± 2.2***
DRB vs. Eb	(38)	47.8 ± 20.8	51.2 ± 8.5[c]	42.4 ± 10.3	11.8 ± 2.4**	54.5 ± 11.5	9.3 ± 2.0***
α chain loci							
monomorphic loci							
DMA vs. Ma	(2)	47.5 ± 26.0	21.4 ± 8.0[a]	79.1 ± 20.0	15.9 ± 3.2**	71.8 ± 14.6	9.7 ± 2.1***
DNA vs. Na	(2)	98.4 ± 53.3	19.8 ± 7.1[a]	59.6 ± 16.2	11.5 ± 3.0**	51.4 ± 11.5	8.2 ± 2.0***
DRA vs. Ea	(2)	37.5 ± 23.8	2.0 ± 2.0	120.5 ± 38.2	16.5 ± 3.6**	60.3 ± 13.3	11.9 ± 2.5***
polymorphic locus							
DQA vs. Aa	(12)	71.0 ± 33.4	29.3 ± 7.2[c]	74.1 ± 18.0	25.6 ± 4.1**	33.0 ± 7.6	15.9 ± 2.9*

Tests of the hypothesis that $d_S = d_N$: *$P < 0.05$; **$P < 0.01$; ***$P < 0.001$. Tests of the hypothesis that d_N equals that for *DRA* versus *Ea* comparison: [a]$P < 0.05$; [b]$P < 0.01$; [c]$P < 0.001$. n, Number of sequences.

chain loci. Here d_N is significantly higher than d_S in the ARS (table 1), which suggests that selection is acting to favour amino acid diversity in this region, as with the polymorphic β chain loci.

4. EVOLUTION OF MONOMORPHIC LOCI

Class I MHC molecules are divided into two subgroups: (1) the class Ia or class I 'classical' loci, which are typically polymorphic and are expressed on almost all cells; and (2) the class of Ib or 'non-classical' class I loci, which are monomorphic or have very low polymorphism and have a much restricted tissue distribution (Klein & Figueroa 1986; Howard 1987). The function of the class Ib loci remains mysterious, although it is now known that at least some of them can have an antigen-presenting function (Pamer *et al.* 1992). Phylogenetic analysis shows that class Ib genes of mammals of one order are more closely related to class Ia genes of the same order than they are to class Ib genes of mammals of other orders (Hughes & Nei 1989*b*). This indicates that class Ib loci have evolved independently in different orders of mammals.

The class II MHC also includes certain monomorphic or nearly monomorphic loci. Certain α chain loci (such as human *DRA*) are monomorphic but form heterodimers with highly polymorphic β chains. In addition, there are also some class II loci that seem more closely analogous with class I 'non-classical' loci in that they are monomorphic and lack a known function. In the mouse the *Na* and *Ob* genes encode a nearly monomorphic heterodimer (Karlsson & Peterson 1992). Homologues of these genes (*DNA* and *DOB*) exist in humans, but in this case it has been suggested that the heterodimer is not expressed because of defects in the *DNA* mRNA (Trowsdale & Kelly 1985). The *DMA* and *DMB* genes (*Ma* and *Mb* in mouse) constitute another pair of genes encoding an almost monomorphic heterodimer (Kelly *et al.* 1991; Cho *et al.* 1991).

To understand phylogenetic relations among polymorphic and monomorphic class II MHC loci in mammals, we constructed a phylogenetic tree based on the conserved Domain 2 amino acid sequences of α and β chain genes (figure 1). Since the gene duplication that gave rise to separate α and β chains occurred early in the history of the MHC, the α chain tree serves to root the β chain tree, and vice versa. As with previous phylogenetic analyses (Hughes & Nei 1990), the tree shows that the different class II regions diverged prior to the divergence of the eutherian orders; thus, for example, *DRB* genes from Primates, Rodentia, Carnivora and Artiodactyla all cluster together, and the cluster is supported by a statistically significant branch (figure 1). Indeed some of these regions may have diverged before the divergence of marsupial (red-necked wallaby) and eutherian mammals. One of two marsupial β chain genes clusters with the *DRB* genes of placental mammals, while the other clusters with *DQB* (figure 1).

The tree shows a contrast between monomorphic class II loci and class Ib loci; unlike the latter, some monomorphic class II loci seem to have diverged early

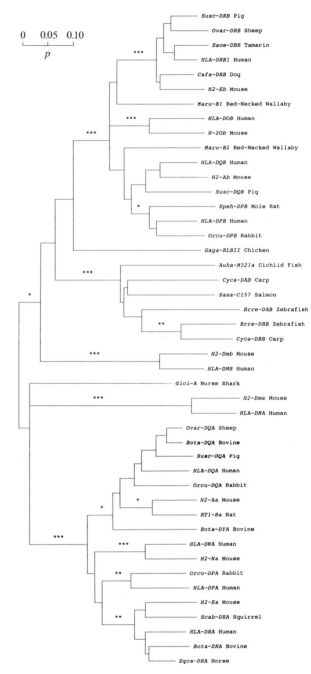

Figure 1. Phylogenetic tree of class II MHC α_2 and β_2 domains, constructed by the minimum evolution method (Rzhetsky & Nei 1992) based on the proportion of amino acid difference (p). Tests of the significance of internal branches: $^*P < 0.05$; $^{**}P < 0.01$; $^{***}P < 0.001$.

in mammalian evolution. Both human *DOB* and *DNA* cluster with their mouse homologues (figure 1). More remarkably, *DMA* clusters outside all other mammalian class II α chain genes, and *DMB* clusters outside all mammalian class II β chain genes. In each case, this pattern is supported by a significant internal branch (figure 1). For class II β chain genes, chicken and bony fish sequences are available, and the *DMB* genes cluster outside both of these, although here the branches are not statistically significant (figure 1). The tree thus suggests that the *DMA–DMB* heterodimer may be a very ancient feature of the vertebrate class II MHC, having arisen by a tandem duplication of α and β chain genes possibly before the divergence of tetrapods from bony fishes.

The *DYA* locus is a non-polymorphic class II α chain locus so far known only from Artiodactyla (Van der Poel *et al.* 1990). This locus clusters with *DQA*, a pattern that is supported by a significant internal branch (figure 1). Thus, this locus appears to have evolved by duplication of a *DQA* gene before radiation of the eutherian orders.

To understand the type of natural selection acting on monomorphic class II loci, we computed d_S and d_N between human and mouse for orthologous pairs of class II genes (table 2). For the polymorphic loci, d_S and d_N in the ARS are about equal in the human–mouse comparison, as observed previously (Hughes & Nei 1988, 1989*a*). This slowdown in the non-synonymous rate relative to the synonymous rate in between-species comparisons is evidently due to a saturation effect at non-synonymous sites (Nei & Hughes 1992). For some monomorphic loci such as *DMB*, d_N in the ARS between human and mouse is nearly as high as that between polymorphic loci. *DOB* and *DRA*, however, show much lower d_N values in the ARS than do the other loci examined. Because the number of synonymous sites in the ARS is small, d_S values have high standard errors; thus, the difference between d_S and d_N in the ARS is significant in none of the human–mouse comparisons; but the fact that d_N in the ARS of *DRA* and *DOB* is significantly lower than that in the other loci (table 2) suggests that this region has been subject to purifying selection in these genes in contrast to the polymorphic loci. Outside the ARS, in comparisons of monomorphic class II genes, d_S is significantly greater than d_N in all comparisons. This indicates that these loci are subject to purifying selection and thus probably encode functional products in human, mouse, or both.

5. POLYMORPHISMS SHARED BETWEEN SPECIES

The existence of long-lasting polymorphisms at MHC loci, leading to sharing of allelic lineages by related species, is now well established (Nei & Rzhetsky 1991). Theoretical studies suggest that there is an upper limit to the amount of time that a polymorphism can be maintained that depends on the effective population size and selection coefficient (Takahata & Nei 1990). Andersson *et al.* (1991) noted that certain polymorphic amino acid residues are shared by DRB molecules of human and bovine. They hypothesized that such sharing of polymorphic residues by mammals of different orders was due to independent evolution, possibly as a result of selection for similar amino acids in the ARS in different species ('convergent evolution').

Lundberg & McDevitt (1992) compared class II β sequences from human and mouse and concluded that shared polymorphism between human and mouse 'represents direct descent of ancestral sequences rather than convergent evolution' (Lundberg & McDevitt 1992, p. 6545). This is a surprising conclusion since primates and rodents probably diverged 80–100 Ma

ago (Li *et al.* 1990). Lundberg & McDevitt found that human and mouse used the same codon at variable β chain positions significantly more frequently than expected on the overall pattern of codon usage found for each species in the database. However, Lundberg & McDevitt did not consider the fact that codon usage in the β_1 domain of human and mouse class II MHC genes is extraordinarily biased. For human and mouse genes in the database (Wada *et al.* 1992), mean third position $G + C$ is 62.0%, whereas that for human and mouse *DQB* (*Ab*) genes is 88.8% and that for human and mouse *DRB* (*Eb*) is 80.8%. When expectations are adjusted for the $G + C$ content bias, there is no significant tendency to share codons between human and mouse or human and bovine (data not shown).

To obtain additional evidence regarding trans-species sharing of polymorphic residues at class II loci, we compared among mammalian species three types of shared amino acid polymorphism.

(1) Shared polymorphism at an *individual residue position* was defined to occur when at least two residues at a position were found to be identical in the two species. For example, at position 9 in the DQ β chain the residues F and Y occur in both human and chimpanzee alleles.

(2) Shared polymorphism at a *two amino acid motif* occurred when, within a sliding window of six aligned residue positions, any two positions were polymorphic in both species and at least two identical amino acid motifs were found in both species. For example, at positions 84 and 85 of the DQ β chain, the sequence motifs EV and QL occur in both human and chimpanzee.

(3) Shared polymorphism at a *three amino acid motif* occurred when, within a sliding window of six aligned residue positions, any two positions were polymorphic in both species and at least two identical amino acid motifs were found in both species. For example, at positions 53, 55 and 57 of the DQ β chain, the sequence motifs QRV and LPA occur in both human and chimpanzee.

The reason for examining these three types of polymorphism was that we expected to find a much lower level of sharing of two and three amino acid motifs when sharing of polymorphism was due to independent evolution rather than to common ancestry. Lundberg & McDevitt note that interallelic recombination may obscure allelic lineages over time, but recombination is unlikely to break up a high proportion of polymorphic motifs within a six amino acid window. Indeed, most evidence on recombination at class I and class II MHC loci suggests that recombination can serve to shuffle polymorphic motifs but not to break them up (Gyllensten *et al.* 1991; She *et al.* 1991; Hughes *et al.* 1993; McAdam *et al.* 1994). For example, putative recombinant alleles at the *DRB1* locus of humans and other Old World primates tend to involve recombination between the β pleated sheet and the α helix of the β chain ARS (Gyllenstein *et al.* 1991). Therefore, it is unlikely that recombination would break up many of the two and three amino acid motifs that we considered.

Table 3. *Sharing of amino acid polymorphisms in β_1 domains of class II MHC β chains by mammalian species*

species (n)	estimated divergence time/Ma	individual positions		2 amino acid motifs		3 amino acid motifs	
		shared	not shared	shared	not shared	shared	not shared
DR β chains							
human (32)							
vs chimpanzee (19)	5–7	32	0	80	0	71	0
vs tamarin (10)	30–40	22	9[c]	14	40[c]	3	38[c]
vs squirrel monkey (11)	30–40	18	16[c]	3	60[c]	0	43[c]
vs dog (9)	60–80	16	4[c]	3	16[c]	0	9[c]
vs bovine (10)	60–80	20	9[c]	6	44[c]	0	40[c]
vs mouse (6)	80–100	13	7[c]	3	18[c]	0	8[c]
DQ β chains							
human (9)							
vs chimpanzee (5)	5–7	12	4	16	3	17	0
vs rhesus monkey (19)	15–25	23	3	22	19	17	17[c]
vs crab-eating macaque (5)	15–25	16	6	16	15[c]	14	6[b]
vs mouse (52)	80–100	13	6	4	20[c]	0	18[c]
vs rat (6)	80–100	8	13	2	16[c]	0	6[c]
chimpanzee (5)							
vs rhesus monkey (19)	15–25	18	0[a]	23	0	19	0
vs crab-eating macaque (5)	15–25	13	4	16	4	15	3
rhesus monkey (19)							
vs crab-eating macaque (5)	1–3	30	0[a]	56	0[b]	49	0
mouse (52)							
vs rat (6)	15–25	12	25[b]	13	23[c]	2	24[c]

Tests of the hypothesis that the proportion shared is the same as that for the human–chimpanzee comparison at the same locus: [a]$P < 0.05$; [b]$P < 0.01$; [c]$P < 0.001$ (Fisher's exact test). n, Number of sequences.

The results (table 3) reveal that there is a remarkably high proportion of shared polymorphic residues between mammalian species. For example, human and chimpanzee shared polymorphisms at every one of the 32 sites in D1 of DR β chains that were polymorphic in both species and at 12 of 16 (75.0%) of such sites in DQ β chains. Even in comparisons between different orders, the proportion of shared residues remains quite high. For example, human and bovine share polymorphisms at 20 of 29 (69.0%) of DR β sites polymorphic in both species, and human and mouse share polymorphisms at 13 of 20 (65.0%) of DR β sites. At the DQB locus, there is not a significant difference between the human–chimpanzee comparison and the human–mouse comparison with respect to the proportion of sites at which there are shared polymorphisms (table 3).

The pattern seen for two and three amino acid motifs is strikingly different. Here, although a high proportion of polymorphism was shared when closely related species such as human and chimpanzee were compared, the proportion of shared polymorphisms was much lower when species of different orders were compared. Indeed, no example of shared polymorphism involving a three amino acid motif between mammals of different orders was found (table 3). Only a very small proportion of three amino acid motif polymorphisms were shared between the human and two New World monkey species, the cottontop tamarin (3 of 41 or 7.3%) and the squirrel monkey (0 of 43) (table 3).

These results are most consistent with the hypothesis that sharing of amino acid poly-morphisms between mammals of different orders involves independent evolution rather than common ancestry. By contrast, in closely related species such as human and chimpanzee, where there is a high proportion of sharing of polymorphisms both for single residues and for two- and three amino acid motifs, common ancestry is the likely explanation for most shared polymorphism, although here also some independent evolution of the same residues probably has occurred as well.

6. CONCLUSIONS

Clearly, MHC polymorphism is maintained by some form of balancing selection that favours diversity in the ARS. It has been proposed that this selection relates to the advantage such diversity confers in enhancing the individual's ability to bind a wide array of foreign peptides and thus to resist a wide array of pathogens (Doherty & Zinkernagel 1975; Hughes & Nei 1988, 1989a). The discovery of a human class I MHC allele and a class II haplotype associated with resistance to *Plasmodium falciparum* malaria in West Africa (Hill *et al.* 1991) provides further support for the hypothesis that selection on MHC loci is driven by pathogens. The fact that natural selection is focused on the antigen-binding portion of the MHC molecule is not consistent with some older hypotheses for MHC polymorphism such as that of mate choice (Thomas 1974) or maternal–fetal incompatibility (Clarke & Kirby 1966).

Overdominant selection can maintain polymorph-isms for millions of years, but maintenance of MHC polymorphisms for over 80 million years, as alleged by

Lundberg & McDevitt, seems unlikely on theoretical grounds (Takahata & Nei 1990). Consistent with this view, our analyses suggest that sharing of polymorphic residues by mammals of different orders is much more likely to be due to independent evolution than to maintenance of an ancestral polymorphism. As regards this issue, the comparisons between DR β chains of human and New World monkeys are particularly instructive. In this case, relatively few two- and three-amino-acid motifs are shared. These species (separated by 30–40 Ma) may be close to the upper limit for sharing of polymorphism through common ancestry. Interestingly, theoretical studies (Takahata & Nei 1990) suggest a maximum of 30–40 Ma for maintenance of MHC polymorphisms in mammals.

This research was supported by grants from the National Institutes of Health to A.L.H. and to M.N. and by a grant from the National Science Foundation to M.N. C.Y.H. is a Howard Hughes scholar.

REFERENCES

Andersson, L., Sigurdardottir, S., Borsch, C. & Gustafsson, K. 1991 Evolution of MHC polymorphism: extensive sharing of polymorphic sequence motifs between human and bovine *DRB* alleles. *Immunogenetics* **33**, 188–193.

Bjorkman, P.J. & Parham, P. 1990 Structure, function, and diversity of class I major histocompatibility complex molecules. *A. Rev. Biochem.* **59**, 253–288.

Bjorkman, P.J., Saper, M.A., Samraoui, B., Bennett, W.S., Strominger, J.L. & Wiley, D.C. 1987*a* Structure of the human class I histocompatibility antigen, HLA-A2. *Nature, Lond.* **329**, 506–512.

Bjorkman, P.J., Saper, M.A., Samraoui, B., Bennett, W.S., Strominger, J.L. & Wiley, D.C. 1987*b* The foreign antigen binding site and T cell recognition regions of class I histocompatibility antigens. *Nature, Lond.* **329**, 512–518.

Brown, J.H., Jardetzky, T., Saper, M.A., Samraoui, B., Bjorkman, P.J. & Wiley, D.C. 1988 A hypothetical model of the foreign antigen binding site of class II histocompatibility molecules. *Nature, Lond.* **332**, 845–850.

Brown, J.H., Jardetzky, T.S., Gorga, J.C., Stern, L.J., Urban, R.G., Strominger, J.L. & Wiley, D.C. 1993 Three dimensional structure of the human class II histocompatibility antigen HLA-DR1. *Nature, Lond.* **364**, 33–39.

Cho, S., Attaya, M. & Monaco, J.J. 1991 New class II-like genes in the mouse MHC. *Nature, Lond.* **353**, 573–576.

Clarke, B. & Kirby, D.R.S. 1966 Maintenance of histocompatibility complex polymorphisms. *Nature, Lond.* **211**, 999–1000.

Doherty, P.C. & Zinkernagel, R. 1975 Enhanced immunologic surveillance in mice heterozygous at the H-2 gene complex. *Nature, Lond.* **256**, 50–52.

Gyllensten, U.B. & Erlich, H.A. 1989 Ancient roots for polymorphism at the HLA–DQα locus in primates. *Proc. natn. Acad. Sci. U.S.A.* **86**, 9986–9990.

Gyllensten, U.B., Sundvall, M. & Erlich, H.A. 1991 Allelic diversity is generated by intraexon sequence exchange at the *DRB1* locus of primates. *Proc. natn. Acad. Sci. U.S.A.* **88**, 3686–3690.

Hill, A.V.S., Allsopp, C.E.M., Kwiatkowski, D. *et al.* 1991 Common West African HLA antigens are associated with protection from severe malaria. *Nature, Lond.* **352**, 595–600.

Howard, J.C. 1987 MHC organization of the rat: evolutionary considerations. In *Evolution and vertebrate immunity* (ed. G. Kelsoe & D. H. Schulze), pp. 397–411. Austin: University of Texas Press.

Hughes, A.L. & Nei, M. 1988 Pattern of nucleotide substitution at major histocompatibility complex class I loci reveals overdominant selection. *Nature, Lond.* **335**, 167–170.

Hughes, A.L. & Nei, M. 1989*a* Nucleotide substitution at major histocompatibility complex class II loci: evidence for overdominant selection. *Proc. natn. Acad. Sci. U.S.A.* **86**, 958–962.

Hughes, A.L. & Nei, M. 1989*b* Evolution of the major histocompatibility complex: independent origin of non-classical class I genes in different groups of mammals. *Molec. Biol. Evol.* **6**, 559–579.

Hughes, A.L. & Nei, M. 1990 Evolutionary relationships of class II major-histocompatibility-complex genes in mammals. *Molec. Biol. Evol.* **7**, 491–514.

Hughes, A.L., Hughes, M.K. & Watkins, D.I. 1993 Contrasting roles of interallelic recombination at the *HLA-A* and *HLA-B* loci. *Genetics* **133**, 669–680.

Karlsson, L. & Peterson, P.A. 1992 The α chain of H-2O has an unexpected location in the major histocompatibility complex. *J. exp. Med.* **176**, 477–483.

Kelly, A.P., Monaco, J.J., Cho, S. & Trowsdale, J. 1991 A new human HLA class II-related locus *DM*. *Nature, Lond.* **353**, 571–576.

Klein, J. 1986 *Natural history of the major histocompatibility complex*. New York: Wiley.

Klein, J. & Figueroa, F. 1986 Evolution of the major histocompatibility complex. *CRC crit. Rev. Immunol.* **6**, 295–386.

Lawlor, D.A., Ward, F.E., Ennis, P.D., Jackson, A.P. & Parham, P. 1988 HLA-A, B polymorphisms predate the divergence of humans and chimpanzees. *Nature, Lond.* **335**, 268–271.

Li, W.-H., Gouy, M., Sharp, P.M., O'hUigin, C. & Yang, Y.W. 1990 Molecular phylogeny of Rodentia, Primates, Artiodactyla, and Carnivora and molecular clocks. *Proc. natn. Acad. Sci. U.S.A.* **87**, 6703–6707.

Lundberg, A.S. & McDevitt, H.O. 1992 Evolution of major histocompatibility complex class II allelic diversity: direct descent in mice and humans. *Proc. natn. Acad. Sci. U.S.A.* **89**, 6545–6549.

McAdam, S.N., Boyson, J.E., Liu, X., Garber, T.L., Hughes, A.L., Bontrop, R.E. & Watkins, D.I. 1994 A uniquely high level of recombination at the *HLA-B* locus. *Proc. natn. Acad. Sci. U.S.A.* **91**, 5893–5897.

Mayer, W.E., Jonker, D., Klein, D., Ivanyi, P., van Seventer, G. & Klein, J. 1988 Nucleotide sequence of chimpanzee MHC class I alleles: evidence for trans-species mode of evolution. *EMBO J.* **7**, 2765–2774.

Nei, M. & Gojobori, T. 1986 Simple methods for estimating the numbers of synonymous and non-synonymous nucleotide substitutions. *Molec. Biol. Evol.* **3**, 418–426.

Nei, M. & Hughes, A.L. 1992 Balanced polymorphism and evolution by the birth-and-death process in the MHC loci. In *Proceedings of the 11th histocompatibility workshop and conference* (ed. K. Tsuji, M. Aizawa & T. Sasazuki), pp. 27–38. Oxford: Oxford University Press.

Nei, M. & Rzhetsky, A. 1991 Reconstruction of phylogenetic trees and evolution of major histocompatibility complex genes. In *Molecular evolution of the major histocompatibility complex* (ed. J. Klein & D. Klein), pp. 13–27. Berlin: Springer-Verlag.

Pamer, E.G., Wang, C.-R., Flaherty, L., Fischer Lindahl, K. & Beran, M.J. 1992 H-2M3 presents a *Listeria monocytogenes* peptide to cytotoxic T lymphocytes. *Cell* **70**, 215–223.

Rzhetsky, A. & Nei, M. 1992 A simple method for estimating and testing minimum evolution trees. *Molec. Biol. Evol.* **9**, 945–967.

She, J.X., Boehme, S.A., Wang, T.W., Bonhomme, F. & Wakeland, E.K. 1991 Amplification of major histocompatibility complex class II gene diversity by intraexonic recombination. *Proc. natn. Acad. Sci. U.S.A.* **88**, 453–457.

Takahata, N. & Nei, M. 1990 Allelic genealogy under overdominant and frequency-dependent selection and polymorphism of major histocompatibility complex loci. *Genetics* **124**, 967–978.

Thomas, L. 1974 Biological signals for self-identification. In *Progress in immunology II* (ed. L. Brent & J. Holborrow), pp. 239–247. Amsterdam: Elsevier.

Trowsdale, J. & Kelly, A. 1985 The human HLA class II α chain gene DZα is distinct from genes in the DP, DQ, and DZ subregions. *EMBO J.* **9**, 2231–2237.

Van der Poel, J.J., Groenen, M.A., Dijkhof, R.J., Ruyter, D. & Giphart, M.J. 1990 The nucleotide sequence of the bovine MHC class II alpha genes: *DRA, DQA,* and *DYA. Immunogenetics* **31**, 29–36.

Wada, K., Wada, Y., Ishibashi, F., Gojobori, T. & Ikemura, T. 1992 Codon usage tabulated from GenBank genetic sequence data. *Nucl. Acids Res.* **20** suppl., 2111–2118.

Discussion

J. C. HOWARD (*Institute for Genetics, University of Cologne, Germany*). In his paper Professor Hughes expressed what he called 'the gene conversion theory' as a form of mutational model for the presence of multiple alleles in the MHC, a model in which allelic frequencies owe their existence to mutational pressure from a conversion-based hypermutator. In this model selection plays no part. This is both old-fashioned and implausible, and not 'the gene conversion theory' which anybody else espouses in this field. I suspect that he has set this crude mutational model up as a straw man, and I for one should be as happy as him to tear it down again.

The gene conversion theory which seems to me to be of interest relates not to the question of allelic frequency (which I take to be regulated by natural and neutral selection in the usual way) but to the nature of the substrates on which selection acts. With selection favouring the rare type, an unusual amount of interest falls on the nature of the process by which rare types are introduced into the population. In such a selective environment, the mutational mechanism which generates the higher overall frequency of successful new alleles will be favoured. If one may distinguish between 'strong' and 'weak' forms of gene conversion theory, the strong asserts that the gene conversion mechanism in the MHC is an adaptive response to the need to generate effective new alleles; the weak form merely eliminates the adaptive component from the argument and asserts that the point mutational mutator exists side by side in the MHC with a gene conversional mutator, perhaps because of accidental features of genomic structure. In the weak form, natural selection may use the products of the gene conversional mutator more effectively, that is, the 'take-up' rate of mutations derived by this mechanism may be higher, but this does not imply that the conversional mutator itself is sustained by natural selection.

If plant gene-for-gene resistance systems also generate new resistance factors by a recombinational mechanism

rather than by point substitution (as seems increasingly likely to be the case, see Ian Crute's paper) I shall consider the strong form of the conversion theory to be highly plausible.

A. L. HUGHES. I have two reasons for doubting the 'strong' version of the gene conversion theory, one empirical and one theoretical. The empirical reason is that we have no evidence that the rate of interlocus recombination ('gene conversion') is particularly high in the MHC. Certainly comparison of DNA sequences has brought to light a number of cases suggestive of past interlocus recombination. The same is true, however, of virtually every multi-gene family for which we have data, and the MHC is not exceptional in this regard. Indeed, if anything, the rate of interlocus recombination may be somewhat lower in the case of the MHC than in the case of other multi-gene families. It is usually assumed that 'gene conversion' involves the formation of a heteroduplex. Because positive selection has acted to diversify MHC genes, the ability to form such a heteroduplex may be reduced in the MHC in comparison to other multi-gene families lacking such diversifying selection.

The theoretical reason is based on arguments by evolutionary biologists that an 'adaptively high' mutation rate will not evolve by natural selection. Most mutations in coding regions are selectively deleterious, and the same is no doubt true of 'gene conversions' as well. Thus, even if a high rate of mutation or 'gene conversion' might be in some way advantageous to the species, it will be disadvantageous to the individual. But natural selection acts primarily at the level of the individual, and thus such a 'group-level' adaptation is not likely to evolve.

In a historical context, it is interesting to note that, in suggesting that there would be a high mutation rate in immune system genes, Haldane relied on a 'group selection' argument. This was at a time when the mechanism by which immunoglobulin diversity is generated was unknown. Using Haldane's argument, one might have predicted that immunoglobulin diversity is generated by a high rate of germline mutation; indeed, many immunologists made exactly this prediction. However, following the logic of natural selection acting at the individual level, one would have predicted rather that somatic mutation or somatic rearrangement of gene segments must be the mechanism, a prediction that has of course turned out to be correct.

P. HIGGS (*Department of Physics, University of Sheffield, U.K.*). How typical is MHC as an example of polymorphic loci in general? In this case the polymorphism has been maintained over a long period. There are presumably many other loci at which new alleles are appearing by mutation all the time and old ones are disappearing due to random drift, thus maintaining a fairly large amount of polymorphism. Is it known what fraction of loci have a long established polymorphism such as MHC? Loci involved in disease resistance may be rather few, but they potentially have a large effect on the fitness of the individuals. If one is interested in population genetics theory with fluctuating selective pressures, as in host–parasite coevolution, one needs to know what are the most important selective pressures acting. Is selection mostly due to a small number of loci with large effects, or due to the sum of a very large number of loci which may each be nearly neutral?

A. L. HUGHES. Among vertebrate genes, the MHC is likely to be unique both in its extraordinarily high level of polymorphism and in its long-lasting polymorphism. Outside the vertebrates, we know of a few loci that parallel these features of the MHC. Some of the loci encoding surface proteins of malaria parasites and the self-incompatibility loci of plants are examples for which a certain amount of sequence data are now available. I am sure that other such examples remain to be discovered, but I expect them to account for only a small proportion of all loci. The type of 'trans-species' polymorphism seen at MHC loci will occur only under balancing selection such as overdominant selection or some type of frequency-dependent selection. However, molecular data suggest that the vast majority of polymorphisms are selectively neutral.

It is somewhat tangential to the question of MHC polymorphism, but my own feeling is that important adaptive breakthroughs generally involve a small number of loci with large effects.

11

The role of infectious disease, inbreeding and mating preferences in maintaining MHC genetic diversity: an experimental test

WAYNE K. POTTS, C. JO MANNING AND EDWARD K. WAKELAND

Center for Mammalian Genetics and Department of Pathology, University of Florida, Gainesville, Florida 32610, and Department of Psychology, University of Washington, Seattle, Washington 98195, U.S.A.

SUMMARY

In house mice, and probably most mammals, major histocompatibility complex (MHC) gene products influence both immune recognition and individual odours in an allele-specific fashion. Although it is generally assumed that some form of pathogen-driven balancing selection is responsible for the unprecedented genetic diversity of MHC genes, the MHC-based mating preferences observed in house mice are sufficient to account for the genetic diversity of MHC genes found in this and other vertebrates. These MHC disassortative mating preferences are completely consistent with the conventional view that pathogen-driven MHC heterozygote advantage operates on MHC genes. This is because such matings preferentially produce MHC-heterozygours progeny, which could enjoy enhanced disease resistance. However, such matings could also function to avoid genome-wide inbreeding. To discriminate between these two hypotheses we measured the fitness consequences of both experimentally manipulated levels of inbreeding and MHC homozygosity and heterozygosity in semi-natural populations of wild-derived house mice. We were able to measure a fitness decline associated with inbreeding, but were unable to detect fitness declines associated with MHC homozygosity. These data suggest that inbreeding avoidance may be the most important function of MHC-based mating preferences and therefore the fundamental selective force diversifying MHC genes in species with such mating patterns. Although controversial, this conclusion is consistent with the majority of the data from the inbreeding and immunological literature.

1. INTRODUCTION

Gene products of the major histocompatibility complex (MHC) play a critical role during immune recognition by serving as antigen receptors that bind peptide fragments for cell-surface presentation to T lymphocytes (Babbitt *et al.* 1985; Bjorkman *et al.* 1987). Each MHC molecule binds a specific subset of peptides (9–20 amino acids in length) representing the intracellular degradation products of both self and non-self proteins. Each T cell expresses a single receptor that is specific for a specific type of MHC–peptide structure. The T-cell receptor repertoire is drawn from a large pool of receptors (estimated at 10^{10}) generated by gene rearrangement processes that are similar to those responsible for antibody diversity (Davis 1985). Before activation, T cells pass through the thymus, where those that recognize MHC–self peptides are terminated, leaving only T cells that recognize MHC–non-self peptides. Because T-cell recognition of an MHC–peptide structure is the triggering event of the immune response, this T-cell selection process confers a primary mode of self–non-self discrimination by the adaptive immune response. The cellular and molecular mechanisms of

this MHC-dependent immune recognition process have been reviewed extensively; see Rotzschke & Falk (1991), Matsumura *et al.* (1992) and references therein.

This crucial function of MHC gene products in immune recognition has led to the widely held view that the unprecedented genetic diversity of MHC genes results from pathogen-driven selection. An MHC-like immune recognition process would be expected to lead to a coevolutionary molecular arms race favouring the genetic diversification of MHC genes. This host–parasite antagonistic coevolutionary process was, in its general form, first predicted by Haldane to account for the extraordinary diversity of vertebrate cell-surface molecules (Haldane 1949). The general hypothesis was later developed by Hamilton (Hamilton 1982; Hamilton *et al.* 1990) and others; see Michod & Levin (1988) for overview papers primarily in the context of the evolutionary forces that maintain sexual reproduction. The adaptation of this process to the diversification of MHC genes was first proposed by Bodmer (1972) and has been continued by many investigators (for example Howard 1991; Slade and McCallum 1992). Pathogens are proposed to evade MHC-dependent immune recognition by mutating their genes encoding MHC-targeted peptides so that they are no longer presented (bound) by MHC, or recognized (bound) by T-cell clones, or both. Such MHC-dependent pathogen evolution directed against a particular MHC phenotype will be undermined when an adapted pathogen infects a conspecific host expressing a different MHC phenotype. Such host individuals will present a different subset of peptides, thereby making previous pathogen evasion events in the context of other MHC phenotypes irrelevant. As a result, this coevolutionary process would be predicted to favour both relatively rare MHC genotypes (negative frequency-dependent selection) and MHC heterozygotes (heterozygote advantage or overdominance) (Doherty & Zinkernagel 1975; Hughes & Nei 1988; Potts and Wakeland 1990; Takahata & Nei 1990; Slade & McCallum 1992; Potts & Wakeland 1993). Both types of selection can, under appropriate conditions, explain observed levels of MHC genetic diversity (Takahata & Nei 1990). The major problem with this elegant and seductive hypothesis is a general lack of empirical confirmation. There are no convincing examples of either MHC heterozygote advantage or MHC-dependent host–parasite coevolution (frequency-dependent selection), although a sufficient amount of circumstantial evidence persists to keep this pathogen-based hypothesis viable (Potts & Wakeland 1993).

If MHC homozygosity is deleterious due to increased susceptibility to infectious disease, then the evolution of reproductive mechanisms that allowed parents to preferentially produce disease-resistant, MHC-heterozygous offspring would be predicted. Mating preferences that accomplish precisely this end have been experimentally demonstrated in *Mus* (house mice), where MHC-dissimilar mates are preferred both under laboratory conditions (Yamazaki *et al.* 1976, 1978, 1988; Egid & Brown 1989) and in semi-natural populations (Potts *et al.* 1991). These MHC-disassortative mating preferences are diversity-maintaining and are sufficient to account for the majority of the genetic diversity observed in *Mus* populations (Potts *et al.* 1991; Hedrick 1992). Furthermore, recent reports of MHC-based mating patterns in humans (Ober *et al.* 1993) suggest that this trait may have some generality in mammals and possibly other vertebrates. The evolution of MHC-based disassortative mating preferences is predicted by the pathogen-driven hypotheses described above, provided there are mechanisms whereby individuals could evaluate the MHC genotypes of prospective mates. Such a mechanism has been convincingly demonstrated in both *Mus* (house mice) (Yamaguchi *et al.* 1981) and *Rattus* (rats) (Singh *et al.* 1987), in that MHC genes influence individual odours in an allele-specific fashion. Mutations in a single MHC gene alter the odour of those individuals carrying the mutation (Yamazaki *et al.* 1983). Thus, the extreme genetic diversity of MHC antigen-binding sites results not only in extensive variation in patterns of antigen presentation, but also in an extensive array of MHC-specific odour types.

An alternative, but not mutually exclusive, function for the evolution of MHC-based disassortative mating preferences is inbreeding avoidance (Brown 1983; Partridge 1988; Uyenoyama 1988; Potts and Wakeland 1990, 1993; Alberts & Ober 1993; Brown & Eklund 1994). The extreme genetic diversity of MHC genes coupled with the olfactory ability to discriminate MHC-mediated odour types by at least some mammals (all those that have been tested, including house mice (Yamaguchi *et al.* 1981), rats (Singh *et al.* 1987) and humans (Gilbert *et al.* 1986)) makes it a potentially useful system for recognizing and avoiding mating with kin (Getz 1981; Potts & Wakeland 1993; Brown 1983; Partridge 1988; Brown & Eklund 1994; Alberts & Ober 1993). For example, by avoiding mating with prospective mates who carry one or more alleles identical to those found in one's own parents, all full- and half-sib matings and half of all cousin matings will be avoided (Potts & Wakeland 1993).

This experimental study was designed to discriminate between the inbreeding and pathogen-mediated heterozygote advantage hypotheses by measuring the fitness consequences of both inbreeding and MHC homozygosity in semi-natural populations of wild-derived *Mus*. In populations we refer to as correlated, MHC-homozygosity was correlated with moderate levels of genome-wide inbreeding, as it generally would be in nature. Animals in uncorrelated populations were systematically bred to eliminate the correlation between MHC homozygosity and inbreeding. We reasoned that if the fitness consequences of inbreeding is more important than MHC homozygosity, then inbred individuals (which are also MHC homozygotes) will have reduced fitness in the correlated populations, whereas the MHC homozygotes in uncorrelated populations will show little or no fitness reduction. Alternatively, if MHC homozygosity is more important, then MHC homozygotes will show reduced fitness in both correlated and uncorrelated populations. If both inbreeding and MHC homozygosity have a similar impact on fitness, then the fitness declines of MHC homozygotes will be greater in correlated populations where both forces are acting in concert.

Perhaps the most important component of fitness for male house mice is their ability to gain and hold territories. The conventional view of the house mouse mating system is that almost all male breeding is done by territory holders (Bronson 1979). Genetic analysis of over 1500 pups analysed from our semi-natural populations by MHC genotyping (a four-allele system) shows that the paternity of all pups was consistent with one of the territorial males, confirming that subordinate males have little or no reproductive success (Potts *et al.* 1992). Consistent with these genetic data, in 41 observed matings, none involved non-territorial males (Potts *et al.* 1991).

Both inbreeding depression and increased disease susceptibility due to MHC homozygosity are expected to negatively affect health and vigour. We predicted that the ability of males to gain and hold territories may be a sensitive measure of health and vigour differences, because males compete and fight aggressively over territories. Consequently, small differences in health and vigour that might be undetectable in non-competitive conditions would be amplified by this head-to-head competition for territories. Thus, we used territoriality as the measure of fitness for males; for females, we used failure or success in reproduction (measured at birth). We are unable to use actual reproductive success as a fitness measure because for males paternity is confounded by high levels of extra-territorial matings by females (Potts *et al.* 1991); for females, maternity is confounded by high levels of communal nesting (Manning *et al.* 1992).

2. METHODS

(a) *Animals*

We had two major requirements for the experimental animals. They must exhibit normal social behaviour and have well-characterized MHC regions. In our experience inbred strains of mice do not exhibit normal social behaviour (Manning *et al.* 1992*b*), whereas crosses between inbred strains and wild-caught mice do (Potts *et al.* 1991; Manning *et al.* 1992*a*). Crossing wild animals with inbred lines allowed us to preserve well-characterized MHC regions from inbred strains while retaining social behaviour described for wild populations (Bronson 1979).

Animals used in these experiments came from generations three and six of original crosses between wild-caught animals and four inbred strains (C57BL/6, BALB/c, B10.BR and DBA/1, carrying MHC haplotypes b, d, k and q, respectively) (figure 1).

In the F$_2$ generation only mice homozygous for one of the four inbred-derived MHC haplotypes were used to continue this outbred colony. In the resulting outbred animals, half of the genome is wild-derived and the other half is derived from one or more inbred strains. The number of inbred strains constituting the inbred-derived portion of the genome was manipulated to test the relative importance of MHC homozygosity and inbreeding (figure 2). In correlated populations, MHC homozygotes were derived from a single inbred strain, whereas MHC heterozygotes were derived from two inbred strains (see figures 1*a* and 2*a*). This established a systematic correlation between inbreeding at MHC loci and genome-wide inbreeding, a condition that would be expected in natural populations that experience some inbreeding (Weir & Cockerham 1973), as is the case for *Mus*. In our breeding design this correlation involved one quarter of the genome and only loci derived from inbred strains. In uncorrelated populations this correlation was eliminated by crosses that systematically introduced uniform contributions of all four inbred strains into each individual (figures 1*b* and 2*b*).

(b) *Physical facility*

Semi-natural populations of mice were housed in a mouse-proof open-air barn at the University of Florida in Gainesville. Ambient temperatures ranged from below 0°C for several days in Dec. 1988 to a high of 39°C in June 1989. The floor was concrete;

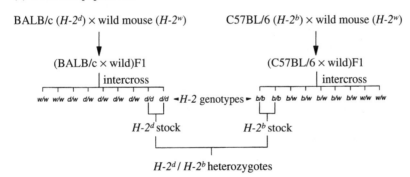

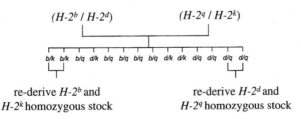

Figure 1. Breeding scheme for the production of founder mice for correlated and uncorrelated populations. (*a*) Correlated populations were produced by breeding wild trapped males with inbred females from BALB/c (*H-2^d*), C57BL/6 (*H-2^b*), DBA/1 (*H-2^q*), and B10.BR (*H-2^k*). The F1 progeny produced in this fashion were intercrossed and F2 progeny homozygous for inbred-derived *H-2* haplotypes were used as founders for *H-2* homozygous strains. *H-2* heterozygotes were produced by crossing appropriate *H-2* homozygotes. This breeding scheme is illustrated with results for BALB/c and C57BL/6. (*b*) Uncorrelated populations were produced by crossing (*H-2^b*/*H-2^d*) heterozygotes with (*H-2^q*/*H-2^k*) heterozygotes and re-deriving *H-2* homozygous stocks.

(a) correlated populations

H-2	remainder of genome
homozygotes	
$H\text{-}2^{d/d}$	50% wild: 50% BALB/c
$H\text{-}2^{b/b}$	50% wild: 50% C57BL/6
$H\text{-}2^{k/k}$	50% wild: 50% B10.BR
$H\text{-}2^{q/q}$	50% wild: 50% DBA/1
heterozygotes	
$H\text{-}2^{b/q}$	50% wild: 25% C57BL/6: 25% DBA/1
$H\text{-}2^{b/d}$	50% wild: 25% C57BL/6: 25% BALB/c
$H\text{-}2^{b/k}$	50% wild: 25% B57BL/6: 25% B10.BR
$H\text{-}2^{d/q}$	50% wild: 25% BALB/c: 25% DBA/1
and so on ...	

(b) uncorrelated populations

all *H-2* homozygotes and heterozygotes	50% wild: 12.5% from each inbred strain

Figure 2. Origins of the genomes of founder mice for correlated and uncorrelated populations. (a) In correlated populations, the relative contribution of various inbred strains to the genomes of *H-2* homozygous stocks differed from that of *H-2* heterozygous stocks, resulting in an increase in genome-wide heterozygosity for *H-2* heterozygotes. (b) In uncorrelated populations, this bias was removed by interbreeding all the stocks and redriving *H-2* homozygous and heterozygous founders.

sidewalls were 0.9 m high sheet metal, which prevented mice from climbing the walls. The remainder of the wall was hardware cloth with 1.25 cm grids, allowing outside air to circulate into the enclosure. The barn was 9.8 m square and was divided in two by a sheet metal barrier 0.8 m high, which allowed two independent populations to proceed simultaneously. Each population had 48 m² of floor space. Each side was divided by hardware cloth 0.4 m high, into eight approximately equal subsections. These barriers did not prevent passage by the mice, but did provide spatial complexity thought to be necessary for normal territorial behaviour (Mackintosh 1970). Additional complexity was provided by a 3 m spiral of hardware cloth 0.4 m high in the centre of each division. Each subsection was provisioned with Purina Rodent Chow, a poultry waterer and five widely spaced nest boxes made from clear plastic one-pint delicatessen containers with a 5 cm hole cut in the side. There was a total of 40 nest boxes and eight food and water stations per population. In addition, a hardware cloth platform 1.5 × 0.4 m was suspended from the ceiling. The platform could be reached from either end by a 0.4 m wide strip of hardware cloth that formed a ramp to the floor. The platform was provisioned with a nest but no food and water. The concrete floors were covered to a depth of 3–4 cm by wood shavings.

(c) Founding stock

All populations were started by releasing 8 males and 16 females into the enclosure. Animals were age-matched and were at least 90 days of age. None had previous sexual experience. Populations were terminated after three to four months, at which point the number of animals (approximately 125) prohibited accurate behavioural observations.

(d) Behavioural observations

During the tenure of the first correlated population (Population A) only data on male–male interactions were taken. For the next six populations (two correlated and four uncorrelated populations) an attempt was made to identify reproductive behaviour of females as well as males. Behavioural observations were made 5–7 times per week for 1–2 h at dusk, and nests were checked nearly every day. Observers worked in pairs, with one or two pairs of observers each evening. Each mouse had a unique combination of notches and holes punched in its ears. This allowed us to identify up to 100 individuals through close-focusing binoculars.

Aggressive interactions, mating behaviour, and obvious pregnancies were noted. Observations of aggressive interactions included information on where chases or fights occurred, which individuals were involved and who was the aggressor (which mouse ran and which chased). Several ten-minute focal studies of each individual were conducted where the location and behaviour of the subject was recorded at 30 s intervals. During daytime nest checks, pups were counted and nursing or attendant females or males were identified. Visibly pregnant females were noted. Records were made of females and/or males sharing sleeping nests with or without litters.

Male territoriality was determined by three criteria. Males were considered to be territorial if they regularly chased other males out of their territories, patrolled and marked boundaries, and were not seen consorting or sleeping with non-territorial subordinate males. Females were classified, in six of the seven populations, as either breeding or non-breeding. Females were considered to be non-breeding if they had not been recorded as pregnant or nursing.

(e) Pathogen load

We attempted to keep the pathogen load in our enclosure populations normal for wild, commensal, house mouse populations in Florida. We trap wild mice and bring them into our colony routinely. Because our colony is in a quarantine facility, we do not practise any special intervention to rid the colony of pathogens or parasites. In limited testing the colony has tested positive for Sendai virus, mouse hepatitis virus, extromelia, polyoma virus and *Mycoplasma pulmonis*.

(f) MHC genotyping

Animals were anaesthetized with metaphane and a 2 cm tail biopsy was taken. The tissue was frozen at −70 °C for later DNA extraction. Restriction fragment length polymorphisms from Taq1-digested genomic DNA allowed identification of all 10 MHC genotypes. Southern blots were hybridized with a 5.8 kb EcoR1 fragment cloned from the mouse MHC class II A$_\beta$ gene. Protocols for DNA extraction, Southern blotting and hybridization are detailed elsewhere (McConnell *et al.* 1988).

3. RESULTS

Three to five males in each population gained and successfully held territories. The remainder of the males became non-territorial subordinates. MHC was the genetic marker we followed for both correlated and uncorrelated populations. If disease-based selection operating through MHC heterozygote advantage was most important then MHC heterozygotes would have an increased probability of gaining territories in both correlated and uncorrelated populations. If inbreeding depression was more important, then MHC heterozygotes would have an advantage in the correlated populations, but the effect would be eliminated in the uncorrelated populations where the correlation between MHC homozygosity and genome-wide inbreeding had been eliminated. Table 1 summarizes the genome-wide inbreeding status of individuals according to MHC genotype and population. Table 2 shows the results of the experiments from three correlated and four uncorrelated populations. MHC heterozygotes enjoyed a fourfold advantage in gaining territories in the correlated population ($p = 0.02$; Fisher Exact Test) but this difference was not observed in the uncorrelated populations.

Table 3 shows that a high percentage of females gave birth in all populations and that among the females that did not give birth there was no correlation with inbreeding or MHC homozygosity.

4. DISCUSSION

In this study we measured the fitness consequences of both experimentally manipulated levels of inbreeding and MHC heterozygosity and homozygosity. We found that neither modest levels of inbreeding nor MHC homozygosity significantly influenced whether a female would give birth or not (Table 3). For males there was a detectable fitness decline associated with inbreeding (Table 2). Relative to more outbred control individuals, inbred males were less likely to gain territories. In contrast, MHC homozygosity had no detectable influence on males fitness (Table 2). These findings suggest that MHC-based disassortative mating

preferences may be more important for avoiding inbreeding than for avoiding MHC homozygosity, at least in *Mus*. Because it appears that MHC genetic diversity is being maintained largely by mating preferences in *Mus* (Potts *et al.* 1991; Hedrick 1992; Potts & Wakeland 1993), these findings indicate that the functional significance of MHC genetic diversity in *Mus* may be associated more with inbreeding avoidance than with resistance to infectious disease. It is important to emphasize that the inbreeding-avoidance and disease-resistance hypotheses are not mutually exclusive. Both inbreeding and disease factors are likely to be operating; the only question is the relative contribution of these two evolutionary forces.

This striking difference in the ability of males to gain territories in the correlated populations must have been due to inbreeding depression and not selection operating directly on MHC genes. The inbreeding load that was experimentally engineered into this system is likely to be less than that found in nature because it only reflects homozygosity for alleles from inbred lines (figure 1). Many of the most deleterious alleles had been purged during the process of creating inbred strains. Consequently, all recessive lethal alleles had been eliminated as well as many of the major deleterious recessive alleles. The inbreeding load analysed in this study represents only a portion of the load in natural populations: the portion that was fixed during the creation of the respective inbred strains. The fact that it strongly affects male territoriality emphasizes the sensitivity of male fitness parameters to inbreeding and probably to any health- and vigour-influencing variable. These data are consistent with the only other studies that measured fitness consequences of inbreeding in an animal system: large inbreeding-associated fitness declines are observed in *Drosophila* (see Charlesworth & Charlesworth (1987) for review).

The inability to detect an effect of inbreeding in

Table 2. *The influence of MHC homozygosity or heterozygosity and inbreeding on the ability of males to gain territorial dominance*

	% males gaining territories	
populations	MHC homozygotes (n)	MHC heterozygotes (n)
correlated	16 (12)	67 (12)
uncorrelated	56 (16)	44 (16)

Table 1. *The genome-wide inbreeding status of MHC homozygotes and heterozygotes in the correlated and uncorrelated populations*

	genome-wide inbreeding status	
populations	MHC homozygotes	MHC heterozygotes
correlated	inbred	outbred
uncorrelated	random bred	random bred

Table 3. *The influence of MHC homozygosity or heterozygosity and inbreeding on the probability of females giving birth*

	% females that gave birth	
populations	MHC homozygotes (n)	MHC heterozygotes (n)
correlated	87 (15)	93 (15)
uncorrelated	77 (30)	78 (32)

females is not surprising for two reasons. First, high variance in male reproductive success but relatively low variance in female reproductive success is the normal condition in most animal systems. In house mice this would be expected because males compete in aggressive encounters for territories, whereas females do not compete directly for breeding rights. Second, whether females did or did not give birth is a relatively crude measure of reproductive success and could have been too insensitive to detect an actual effect. Lifetime reproductive success would be a better measure; we plan to genetically determine parentage in these populations to make this evaluation.

The experimental design of this study has two weaknesses. First, the inbreeding load was probably not normal. However, as argued above, it was probably weaker than normal; thus the main conclusion of this paper is conservative for the inbreeding effects found for males. Second, we cannot ensure that the parasites and pathogens present in our population represent normal levels found in natural populations. As described above, we attempt to keep normal pathogen loads in our enclosure populations; but because many, if not most, *Mus* pathogens have not yet been characterized this is impossible to document. Consequently, the inability to detect fitness declines due to MHC homozygosity may be due in part to an incomplete representation of *Mus* pathogens in the enclosure populations.

The data from this study suggest that, at least in *Mus*, avoiding inbreeding plays an important role in the evolution of MHC-based mating preferences, and consequently the evolution of MHC genetic diversity. Although this conclusion differs substantially from the normal immunological view, it is quite consistent with a careful review of the literature relevant to inbreeding and MHC-related resistance to infectious disease. The general conclusions in the literature are that inbreeding depression is a major fitness-reducing force that is found in essentially all species investigated, and that many species have genetic-based inbreeding-avoidance systems, often based on a single highly polymorphic locus (Charlesworth & Charlesworth 1987). A similar level of inbreeding depression in *Mus* would be predicted to drive the evolution of genetic-based disassortative mating preferences and as a consequence would favour MHC diversity, owing to MHC's central role in influencing odour types (Yamaguchi *et al.* 1981). In contrast, MHC-hetero-zygote advantage has never been convincingly demonstrated and MHC-related resistance to infectious disease has not been found in most cases investigated (Tiwari & Terasaki 1985; Klein 1986). The two leading exceptions to this – malaria in humans (Hill *et al.* 1991) and Marek's disease in domestic chickens (Pazderka *et al.* 1975) – are the wrong kind of disease associations because they strongly reduce fitness and they favour only one or two MHC alleles, thus resulting in reduced diversity.

In the following sections we provide brief entries to the literature on inbreeding depression, inbreeding avoidance, MHC as a kin-recognition marker, and pathogen-driven selection operating on MHC genes. Most of these areas have been recently reviewed. We take this opportunity to point out very recent findings and important ideas that are often overlooked and omitted from such discussions.

(a) Inbreeding depression, inbreeding avoidance and MHC-based kin recognition

The capacity of MHC polymorphisms to generate strong olfactory signals may have resulted in its use to avoid inbreeding through an olfaction-based kin-recognition system. The well-documented MHC-based olfactory signals of rodents provide MHC-genotype-specific information (Yamazaki *et al.* 1979). Although the molecular mechanisms responsible for the genera-tion of these odorants are unknown, a variety of studies have directly correlated changes in odorants with specific mutations in the antigen-binding sites of MHC class I and class II molecules (Boyse *et al.* 1987). Therefore, the extensive diversity of MHC-binding sites results not only in extensive variation in the antigens presented, but also in an extensive array of odour types that are MHC-genotype-specific. Consequently, MHC-mediated odours provide genetic-based kin recognition information and disassortative MHC-based mating preferences will not only result in the preferential production of MHC-heterozygotes, but will also favour outbreeding and an increase in genome-wide heterozygosity (Uyenoyama 1988). This effect will be especially strong in small populations where prospective mating partners include many related individuals, as in *Mus* populations (Hartl & Clark 1989). Enhanced fitness due to increased genome-wide heterozygosity or inbreeding avoidance may strongly favour the evolution of MHC-based mating preferences.

In addition to this study, two other lines of evidence support the importance of inbreeding avoidance in the evolution of MHC-based mating preferences. The first is an independent demonstration that house mice, presumably through their demonstrated ability to detect MHC-genotyoe-specific odours (Yamazaki *et al.* 1979), use MHC for genetic-based kin recognition (Manning *et al.* 1992a). *Mus* is a communally nesting species that displays the relatively rare mammalian trait of (apparent) indiscriminate communal nursing. Such behaviour is subject to kin selection, which would promote communal nesting among kin. Restricting shared nursing to kin would decrease the probability of exploitation (Grafen 1990) and increase inclusive fitness. If MHC genes are being used as a kin recognition marker, then communal nesting partners are predicted to have high MHC similarity. When familiar sisters were available, they were strongly preferred as communal nesting parterns. However, when sisters were not available or this variable was controlled statistically, MHC-similar females were preferred as communal nesting parterns (Manning *et al.* 1992a). The most plausible interpretation of these data is that MHC genes are being used as kin-recognition markers to increase relatedness among communal nesting partners. These data represent the first unambiguous demonstration of genetic-based kin recognition in any vertebrate species and consequently provide support for the inbreeding avoidance hypothesis. Such kin-recognition genes have been predicted and sought for some time (Hamilton 1964).

The second line of evidence is the observation that inbreeding depression is a strong and pervasive phenomenon of living systems (Charlesworth & Charlesworth 1987) and that many specific mechanisms have evolved throughout the plant and animal kingdoms to avoid inbreeding (Uyenoyama 1988; Blouin & Blouin 1988). Consistent with an MHC-based inbreeding avoidance system, we have recently demonstrated that the only other genetic system known to display all of the unique genetic features of MHC genes – an extreme number of alleles, ancient allelic lineages that pre-date contemporary species (trans-species evolution), extremely high sequence divergence of alleles, and high rates of non-synonymous substitutions – are plant self-incompatibility genes (Potts & Wakeland 1993). These data demonstrate empirically that all of the extreme and unusual features of MHC genetic diversity, features that are routinely cited as evidence for pathogen-drived selection, could be due to mating patterns. Disassortative mating patterns in plants have led to the only other example of such extreme genetic features and it is generally agreed that plant self-incompatibility systems evolved to reduce inbreeding.

This study and these lines of evidence suggest that, at least for *Mus*, inbreeding avoidance may be an important force leading to both the evolution of MHC-based mating preferences and the maintenance of MHC genetic diversity. MHC-based kin recognition is likely to occur in species with MHC-mediated odours, which is probably most mammals (see Brown & Eklund (1994) for recent review). MHC-mediated odour recognition in other vertebrates is perhaps less likely on grounds that the sense of smell is less acutely developed in non-mammalian vertebrates, but appropriate studies have not been done.

(b) Pathogen-driven balancing selection operating on MHC genes

We and others have reviewed the evidence for pathogen-driven balancing selection (Kiltz *et al.* 1984; Tiwari & Terasaki 1985; Klein 1986; Potts & Wakeland 1990; Takahata & Nei 1990; Slade & McCallum 1992; Hughes & Nei 1992; Potts & Wakeland 1993). The most important point to emerge is that the two types of data that could directly document pathogen-driven balancing selection – pathogen evolution in response to MHC-dependent immune recognition or MHC heterozygote advantage due to infectious disease – have not been unambiguously documented. Given this point, it is quite easy to look at the remainder of the indirect and circumstantial data and argue for or against the importance of pathogen-driven selection. For example, the most important indirect data –

increased susceptibility to infectious disease based on MHC genotype – although not a prevalent feature of host–parasite interactions, has been detected (Pazderka *et al.* 1975; Lynch *et al.* 1986; Wassom *et al.* 1987; Hirayama *et al.* 1988; Nauciel *et al.* 1988; Hormaeche *et al.* 1985; Kennedy *et al.* 1986; Hamelin-Bourassa *et al.* 1989; Hill *et al.* 1991). Some view this as a meagre listing that represents a crippling blow to the pathogen-driven hypothesis; others see these studies as adequate evidence for pathogen-driven selection. We believe it is premature for any major conclusions and that new data will be required to evaluate the relative importance of pathogen-driven balancing selection operating on MHC genes. The following four observations or ideas are relatively new and are missing from most of the above reviews; they should be added to the debate.

First, there have been two recent observations that are consistent with a model of pathogen evasion of MHC-dependent immune recognition. High levels of replacement substitutions have been observed in a region of HIV (the human innunodeficiency virus) known to be used in antigen presentation by the specific MHC of the host individual from which the virus was collected (Phillips *et al.* 1991). This same region showed low substitution frequencies in virus collected from individuals with a different MHC. These data are consistent with the possibility that viral variants were favoured because they escaped MHC-based immune recognition at this epitope. Similarly, human populations in New Guinea, where there is a high incidence of human leukocyte antigen (HLA) A-11, are commonly infected with a variant of the Epstein–Barr virus which has a replacement substitution in an epitope presented by HLA A-11 (De Campos *et al.* 1993). This variant epitope is no longer presented by HLA A-11 individuals, thus suggesting that this viral variant is common because it has an advantage in HLA A-11 individuals. These two observations are intriguing and should be followed by experiments designed to test for pathogen evasion of MHC-dependent immune recognition.

Second, pathogen evasion of MHC-dependent immune recognition as the driving force behind MHC genetic diversity is rarely scrutinized. One idea that has been missing from the debate is the following possibility. Even pathogens with small genomes express many MHC-presented T-cell epitopes and each epitope has many T-cells that recognize this MHC–peptide in different ways. Consequently, it may be difficult or even impossible for any pathogen to accumulate all the necessary mutations to totally evade MHC-dependent immune recognition of all epitopes. If evasion of some but not all epitopes has little consequence for the ability of the host to respond effectively to the pathogen, then pathogen evasion will be a weak force in the evolution of MHC genetic diversity. If one were to design an immune system, a primary goal would be to make it difficult or impossible for pathogens to evade; perhaps evolution has achieved this. Again, experiments designed to test for pathogen evasion of MHC-dependent immune recognition are needed.

Third, when MHC class I-deficient mice were made, it was anticipated that they would have major deficits in their adaptive immune response and that they would be extremely susceptible to infectious disease. The big surprise is that these mice appear quite healthy in colony conditions and have shown either no or relatively minor increases in susceptibility to experimental infections (Koller and Smithies 1989; Koller *et al.* 1990; Eichelberger *et al.* 1991; Bender *et al.* 1992; Muller *et al.* 1992; Spriggs *et al.* 1992; Epstein *et al.* 1993; Grusby *et al.* 1993; Roberts *et al.* 1993). If it is difficult to document important disease susceptibilities in MHC class I-deficient mice, the task of documenting differential disease susceptibilities between MHC homozygotes and heterozygotes will be far more difficult.

Fourth, if pathogen-driven heterozygous advantage is as strong as either the inbreeding depression measured in many species (Charlesworth & Charlesworth 1987) or the MHC-based mating preferences measured in seminatural populations of house mice (Potts *et al.* 1991), then it would have been detected by now at least in humans and house mice, where there has been extensive work on MHC genes and infectious disease. This suggest that in species with MHC-based mating preferences similar to those observed in *Mus*, these mating preferences may be the primary force maintaining MHC genetic diversity and they will function primarily to avoid inbreeding.

This work was conducted while E.K.W. was supported by a NIH grant and while W.K.P. was supported by grants from NIH and NSF.

REFERENCES

Alberts, S.C. & Ober, C. 1993 Genetic variability in the major histocompatibility complex: A review of non-pathogen-mediated selective mechanisms. *Phys. Anthropol.* **36**, 71–89.

Babbitt, B.P., Allen, P.M. & Matsueda, G. 1985 Binding of immunogenic peptides to Ia histocompatible molecules. *Nature, London* **317**, 359–361.

Bender, B.S., Croghan, T., Zhang, L. & Small, P.A. 1992 Transgenic mice lacking class I major histocompatibility complex-restricted T cells have delayed viral clearance and increased mortality after influenza virus challenge. *J. exp. Med.* **175**, 1143–1145.

Bjorkman, P.J., Saper, M.A., Samraoui, B., Bennett, W.S., Strominger, J.L. & Wiley, D.C. 1987 The foreign antigen binding site and T cell recognition regions of class I histocompatibility antigens. *Nature, Lond.* **329**, 512–518.

Blouin, S.F. & Blouin, M. 1988 Inbreeding avoidance behaviors. *Trends Ecol. Evol.* **3**, 230–233.

Bodmer, W.F. 1972 Evolutionary significance of the HLA system. *Nature, Lond.* **237**, 139–145.

Boyse, E.A., Beauchamp, G.K. & Yamazaki, K. 1987 The genetics of body scent. *Trends Genet.* **3,** 97–102.

Bronson, F.H. 1979 The reproductive ecology of the house mouse. *Q. Rev. Biol.* **54**, 265–299.

Brown, J.L. 1983 Some paradoxical goals of cells and organisms: the role of the MHC. In *Ethical questions in brain and behaviour* (ed. D. W. Pfaff), pp. 111–124. New York: Springer-Verlag.

Brown, J.L. & Eklund. A. 1994 Kin recognition and the major histocompatibility complex: An integrative review. *Am. Nat.* **143**, 435–461.

Charlesworth, D. & Charlesworth, B. 1987 Inbreeding depression and its evolutionary consequences. *A. Rev. Ecol. Syst.* **18**, 237–268.

Davis, M.M. 1985 Molecular genetics of the T cell-receptor beta chain. *A. Rev. Immunol.* **3**, 537–560.

De Campos, P.O., Gavioli, R., Zhang, Q.J. *et al.* 1993 HLA-A11 epitope loss isolates of Epstein-Barr virus from a highly A11+ population. *Science, Wash* **260**, 98–100.

Doherty, P.C. & Zinkernagel, R.M. 1975 Enhnaced immunological surveillance in mice heterozygous at the H-2 gene complex. *Nature, Lond.* **256**, 50–52.

Egid, K. & Brown, J.L. 1989 The major histocompatibility complex and female mating preferences in mice. *Anim. Behav.* **38**, 548–549.

Eichelberger, M., Allan, W., Zijlstra, M., Jaenisch, R. & Doherty, R.C. 1991 Clearance of influenza virus respiratory infection in mice lacking class I major histocompatibility complex-restricted CD8+ T cells. *J. exp. Med.* **174**, 875.

Epstein, S.L., Misplon, J.A., Lawson, C.M., Subbarao, E.K., Connors, M. & Murphy, B.R. 1993 Beta 2-microglobulin-deficient mice can be protected against influenza A infection by vaccination with vaccinia-influenza recombinants expressing hemagglutinin and neuraminidase. *J. Immunol.* **150**, 5484–5493.

Getz, W.M. 1981 Genetically based kin recognition systems. *J. theor. Biol.* **92**, 209–226.

Gilbert, A.N., Yamazaki, K. & Beauchamp, G.K. 1986 Olfactory discrimination of mouse strains (*Mus musculus*) and major histocompatibility types by humans (*Homo sapiens*). *J. comp. physchol.* **100**, 262–265.

Grafen, A. 1990 Do animals really recognize kin? *Anim. Behav.* **39**, 42–54.

Grusby, M.J., Auchincloss, H., Lee, R. *et al.* 1993 Mice lacking major histocompatibility complex class I and II molecules. *Proc. natn. Acad. Sci. U.S.A.* **90**, 3913–3917.

Haldane, J.B.S. 1949 Disease and evolution. *Ricerca scient.* **19**, 68–75.

Hamelin-Bourassa, D., Skamene, E. & Gervais, F. 1989 Susceptibility to mouse acquired immunodeficiency syndrome is influenced by the H-2. *Immunogenetics* **30**, 266–272.

Hamilton, W.D. 1964 The genetical evolution of social behaviour. *J. theor. Biol.* **7**, 1–52.

Hamilton, W.D. 1982 Pathogens as causes of genetic diversity in their host populations. In: *Population biology of infectious diseases* (ed. D. Konferenzen, R. M. Anderson & R. M. May), pp. 269–296. Heidelberg: Springer-Verlag.

Hamilton, W.D., Axelrod, R. & Tanese, R. 1990 Sexual reproduction as an adaption to resist parasites (a review). *Proc. natn. Acad. Sci. U.S.A.* **87**, 3566–3573.

Hartl, D.L. & Clark, A.G. (1989) *Principles of population genetics.* Sunderland, Massachusetts: Sinauer Associates.

Hedrick, P.W. 1992 Female choice and variation in the Major Histocompatibility Complex. *Genetics* **132**, 575–581.

Hill, A.V.S., Allsop, C.E.M., Kwiatkowski, D. *et al.* 1991 Common West African HLA antigens are associated with protection from severe malaria. *Nature, Lond.* **352**, 595–600.

Hirayama, K., Matsushita, S., Kikuchi, I., Iuchi, M., Ohta, N. & Sasazu, T. 1988 HLA-DQ is epistatic to HLA-DR in controlling the immune response to schistosomal antigen in humans. *Nature, Lond.* **327**, 426–429.

Hormaeche, C.E., Harrington, K.A. & Joysey, H.S. 1985 Natural resistance to salmonellae in mice: control by genes within the major histocompatibility complex. *J. infect Dis.* **152**, 1050–1056.

Howard, J.C. 1991 Disease and evolution. *Nature, Lond.* **352**, 565–566.

Hughes, A.L. & Nei, M. 1988 Pattern of nucleotide substitution at major histocompatibility complex class I loci reveals overdominant selection. *Nature, Lond.* **335**, 167–170.

Hughes, A.L. & Nei, M. 1992 Models of host-parasite interaction and MHC polymorphism. *Genetics* **132**, 863–864.

Kennedy, M.W., Gordon, A.M.S., Tomlinson, L.A. & Qureshi, F. 1986 Genetic (major histocompatibility complex?) control of the antibody repertoire to the secreted antigens of *Ascaris*. *Parasit. Immunol.* **9**, 269–273.

Klein, J. 1986 *Natural history of the major histocompatibility complex.* New York: Wiley.

Klitz, W., Thomson, G. & Baur, M.P. 1984 The nature of selection in the HLA region based on population data from the ninth workshop. In *Histocompatibility testing 1984* (ed. E. D. Albert) pp. 330–332. Berlin: Springer-Verlag.

Koller, B.H., Marrack, P., Kappler, J.H. & Smithies, O. 1990 Normal development of mice deficient in B_2, MHC class I proteins, and CD8$^+$ T cells. *Science, Wash.* **248**, 1227–1230.

Koller, B.H. & Smithies, O. 1989 Inactivating the β_2-microglobulin locus in mouse embryonic stem cells by homologous recombination. *Proc. natn. Acad. Sci. U.S.A.* **86**, 8932–8935.

Lynch, D.H., Cole, B.C., Bluestone, J.A. & Hodes, R.J. 1986 Cross-reactive recognition by antigen-specific, major histocompatibility complex-restricted T cells of a mitogen derived from *Mycoplasma arthritides* is clonally expressed and I-E restricted. *Eur. J. Immunol.* **16**, 747–751.

Mackintosh, J.H. 1970 Territory formation by laboratory mice. *Anim. Behav.* **18**, 177–183.

Manning, C.J., Wakeland, E.K. & Potts, W.K. 1992*a* Communal nesting patterns in mice implicate MHC genes in kin recognition. *Nature, Lond.* **360**, 581–583.

Manning, C.J., Potts, W.K., Wakeland, E.K. & Dewsbury, D.A. 1992*b* What's wrong with MHC mate choice experiments? In *Chemical signals in vertebrates*, vol. 6 (ed. R. L. Doty & D. Muller-Schwarze), pp. 229–235. New York: Plenum Press.

Matsumura, M., Fremont, D.H., Peterson, P.A. & Wilson, I.A. 1992 Emerging principles for the recognition of peptide antigens by MHC class I molecules. *Science, Wash.* **257**, 927–934.

McConnell, T.J., Talbot, W.S., McIndoe, R.A. & Wakeland, E.K. 1988 The origin of MHC class II gene polymorphism within the genus Mus. *Nature, Lond.* **332**, 651–654.

Michod, R.E. & Levin, B.R. 1988 *The evolution of sex.* Sunderland, Massachusetts: Sinauer Associates.

Muller, D., Koller, B.H., Whitton, J.L., LaPan, K.E., Brigman, K.K. & Frelinger, J.A. 1992 LCMV-specific, class II-restricted cytotoxic T cells in beta 2-microglobulin-deficient mice. *Science, Wash.* **255**, 1576–1578.

Nauciel, C., Ronco, E., Guenet, J.L. & Marika, P. 1988 Role of H-2 and non-H-2 genes in control of bacterial clearance from the spleen in *Salmonella typhimurium*-infected mice. *Infect. Immun.* **56**, 2407–2411.

Ober, C., Weitkamp, L.R., Elias, S. & Kostyu, D.D. 1993 Maternally-inherited HLA haplotypes influence mate choice in a human isolate. *Hum. Genet.* **53**, 206. (Abstract.)

Partridge, L. 1988 The rare-male effect: what is its evolutionary significance? *Phil. Trans. R. Soc. Lond.* B **319**, 525–539.

Pazderka, F., Longenecker, B.M., Law, G.R.J., Stone, H.A. & Ruth, R.F. 1975 Histocompatibility of chicken populations selected for resistance to Marek's disease *Immunogenetics* **2**, 93–100.

Phillips, R., Rowland-Jones, S., Nixon, F.D. *et al.* 1991 Human immunodeficiency virus genetic variation that can escape cytotoxic T cell recognition. *Nature, Lond.* **354**, 453.

Potts, W.K., Manning, C.J. & Wakeland, E.K. 1991 Mhc genotype influences mating patterns in semi-natural populations of *Mus*. *Nature, Lond.* **352**, 619–621.

Potts, W.K., Manning, C.J. & Wakeland, E.K. 1992 MHC-based mating preferences in *Mus* operate through both settlement patterns and female controlled extra-territorial matings. In *Chemical signals in vertebrates*, vol. 6 (ed. R. L. Doty & D. Muller-Schwarze), pp. 229–235. New York: Plenum Press.

Potts, W.K. & Wakeland, E.K. 1990 Evolution of diversity at the major histocompatibility complex. *Trends Ecol. Evol.* **5**, 181–187.

Potts, W.K. & Wakeland, E.K. 1993 The evolution of MHC genetic diversity: a tale of incest, pestilence and sexual preference. *Trends Genet.* **9**, 408–412.

Roberts, A.D., Ordway, D.J. & Orme, I.M. 1993 *Listeria monocytogenes* infection in beta 2 microglobulin-deficient mice. *Infect. Immun* **61**, 1113–1116.

Rotzschke, O. & Falk, K. 1991 Naturally-occurring peptide antigens derived from the MHC class-I-restricted processing pathway. *Immunol. Today* **12**, 447–455.

Singh, P.B., Brown, R.E. & Roser, B. 1987 MHC antigens in urine as olfactory recognition cues. *Nature, Lond.* **327**, 161–164.

Slade, R.W. & McCallum, H.I. 1992 Overdominant *vs.* frequency-dependent selection at MHC loci. *Genetics* **132**, 861–862.

Spriggs, M.K., Koller, B.H., Sato, T. *et al.* 1992 Beta 2-microglobulin, CD8+ T-cell deficient mice survive inoculation with high doses of vaccinia virus and exhibit altered IgG responses. *Proc. natn. Acad. Sci. U.S.A.* **89**, 6070–6074.

Takahata, N. & Nei, M. 1990 Allelic genealogy under overdominant and frequency-dependent selection and polymorphism of Major Histocompatibility Complex loci. *Genetics* **124**, 967–978.

Tiwari, J.L. & Terasaki, P.I. 1985 *HLA and disease associations.* New York: Springer-Verlag.

Uyenoyama, M.K. 1988 On the evolution of genetic incompatibility systems: incompatibility as a mechanism for the regulation of outcrossing distance. In The evolution of sex (ed. R. E. Michod & B. R. Levin), pp. 212–232. Sunderland, Massachusetts: Sinauer Associates.

Wassom, D.L., Krco, C.J. & David, C.S. 1987 I-E expression and susceptibility to parasite infection. *Immunol. Today* **8**, 39–43.

Weir, B.S. & Cockerham, C.C. 1973 Mixed self and random mating at two loci. *Genet. Res.* **21**, 247–262.

Yamaguchi, M., Yamazaki, K., Beauchamp, G.K., Bard, J., Thomas, L. & Boyse, E.A. 1981 Distinctive urinary odors governed by the major histocompatibility locus of the mouse. *Proc. natn. Acad. Sci. U.S.A.* **78**, 5817.

Yamazaki, K., Boyse, E.A., Mike, V. *et al.* 1976 Control of mating preferences in mice by genes in the major histocompatibility complex. *J. exp. Med.* **144**, 1324–1335.

Yamazaki, K., Yamaguchi, M., Andrews, P.W., Peake, B. & Boyse, E.A. 1978 Mating preferences of F^2 segregants of crosses between MHC-congenic mouse strains. *Immunogenetics* **6**, 253–259.

Yamazaki, K., Yamaguchi, M., Baranoski, L., Bard, J., Boyse, E.A. & Thomas, L. 1979 Recognition among mice: Evidence from the use of a Y-maze differentially scented by congenic mice of different major histocompatibility types. *J. exp. Med.* **150**, 755–760.

Yamazaki, K., Beauchamp, G.K., Egorov, I.K., Bard, J., Thomas, L. & Boyse, E.A. 1983 Sensory distinction between H-2^b and H-2^{bm1} mutant mice. *Proc. natn. Acad. Sci. U.S.A.* **80**, 5685–5688.

Yamazaki, K., Beauchamp, G.K., Kupniewski, D., Bard, J., Thomas, L. & Boyse, E.A. 1988 Familial imprinting determines H-2 selective mating preferences. *Science, Wash.* **240**, 1331–1332.

Discussion

A. L. HUGHES (*Department of Biology, Pennsylvania State University, U.S.A.*). Dr Potts' experiment failed to discriminate between two hypotheses: (i) mice mate disassortatively based on MHC; and (ii) mice avoid mating with their siblings, which they recognize by a variety of cues including MHC-based ones.

The second of these will not enhance polymorphism in a large natural population. To discriminate between these, one needs to conduct additional experiments. I will suggest one. A situation could be set up in which females can choose among the following: (a) a familiar full sibling, which shares neither MHC haplotype with the female: (b) an unfamiliar unrelated male which shares one or both of its MHC haplotypes with the female; and (c) an unfamiliar unrelated male which shares neither MHC haplotype with the female.

On the hypothesis of MHC-based dissassortative mating, females should prefer (a) and (c) to (b), but should not discriminate between (a) and (c). If females only avoid sib-mating, they should prefer (b) and (c) to (a) but should not discriminate between (b) and (c). If females avoid sib-mating but also mate dissociatively based on the MHC, their preference should be in the order (c) > (b) > (a).

Further experiments could address the roles of familiarity and genotype in the recognition, which Dr Potts has also failed to address.

W. K. POTTS. Dr Hughes' two hypotheses do not seem to be alternatives to each other. The first hypothesis states that mice have MHC disassortative mating preferences: the second states that mice have MHC disassortative mating preferences as well as other mechanisms to avoid sib matings. From the mating preference study suggested by Dr Hughes it seems that the question of interest is: Are familiarity or MHC-based cues more important in mating preferences? Our original study (Potts *et al.* 1991) included Dr Hughes' conditions (b) and (c); females in semi-natural populations were allowed to choose between unfamiliar unrelated males that were either MHC similar or dissimilar. They preferred MHC dissimilar males indicating that MHC-based mating preferences are used in situations not involving sibs. We avoided offering females familiar sib males due to the expectation that these males would be avoided independent of MHC cues. We agree with Dr Hughes that this is an assumption in need of testing. But we disagree with Dr Hughes when he says that MHC disassortative mating preferences that function to avoid sib matings will not enhance polymorphism in a large natural population. It has been shown that genetic-based disassortative mating preferences have similar population genetic dynamics to heterozygote advantage (Karlin, S. & Feldman, M.W. 1968 Further analysis of negative assortive mating. *Genetics* **59**, 117–136) and will be a strong selective force favouring polymorphism, independent of whether they function in conjunction with other cues to avoid sib matings. For example, plant self-incompatibility loci which function in conjunction with other mechanisms to avoid matings with self, sibs and other close relatives display extreme genetic diversity (Potts & Wakeland 1993).

12

Human leukocyte antigens and natural selection by malaria

ADRIAN V. S. HILL[1], SIMON N. R.YATES[1], CATHERINE E. M. ALLSOPP[1], SUNETRA GUPTA[2], SARAH C. GILBERT[1], AJIT LALVANI[1], MICHAEL AIDOO[1], MILES DAVENPORT[1] AND MAGDALENA PLEBANSKI[1]

[1] Molecular Immunology Group, Institute of Molecular Medicine, University of Oxford, John Radcliffe Hospital, Oxford OX3 9DU, U.K.
[2] Department of Zoology, University of Oxford, South Parks Road, Oxford OX1 3PS, U.K.

SUMMARY

The extraordinary polymorphism of human leukocyte antigens (HLA) poses a question as to how this remarkable diversity arose and is maintained. The explanation that infectious pathogens are largely responsible is theoretically attractive but clear and consistent associations between HLA alleles and major infectious diseases have rarely been identified. Large case-control studies of HLA types in African children with severe malaria indicate that HLA associations with this parasitic infection do exist and it is becoming possible to investigate the underlying mechanisms by identification of peptide epitopes in parasite antigens. Such analysis reveals how the magnitude and detectability of HLA associations may be influenced by numerous genetic and environmental factors. These complex interactions will give rise to variation over time and space in the selective pressures exerted by infectious diseases and this fluctuation may, in itself, contribute to the maintenance of HLA polymorphism.

1. INTRODUCTION

The major histocompatibility complex (MHC) encodes many of the most polymorphic genes in humans. Those in the class I region encode the highly variable HLA-A, -B and -C proteins which play a central role in the activation of cytotoxic T lymphocytes by presenting short peptides of cytoplasmic origin to these cells. In turn, once activated, these CTL are capable of lysing and killing cells presenting the same antigenic peptide. Hence, this system appears particularly well adapted to defence against intracellular pathogens.

In contrast, the class II region of the MHC encodes HLA-DR, -DQ and -DP gene products which assemble to produce HLA class II molecules that present somewhat longer peptides (10–25 amino acids) to helper T lymphocytes bearing the surface molecule CD4. These helper T cells assist B cells in antibody production and can release a range of cytokines which activate macrophages and other cells. HLA class II molecules are therefore of central importance in regulating immune defence against extracellular as well as some intracellular pathogens.

The pivotal role that MHC molecules play in immune responses against infectious pathogens led to the proposal that the remarkable diversity of MHC molecules had evolved as a result of such host–parasite interactions (Snell 1968; Doherty & Zinkernagel 1975). However, for many years evidence on this point has been controversial. In particular, studies of HLA–disease associations in humans revealed more convincing associations with autoimmune disease than with any infection. In this short review we describe our studies of HLA and malaria in African populations and set these in the context of new molecular insights into the mechanisms of HLA associations with disease in humans. These recent studies help to clarify some of the uncertainties in the older literature and support a major role for pathogen-driven selection in maintaining MHC diversity in human populations.

2. HLA ASSOCIATIONS WITH INFECTIOUS DISEASES

Many studies of HLA associations with infectious diseases have been reported over the last twenty years. The early studies were often very small and could only consider the limited number of HLA alleles known at the time. Clear associations were seldom seen and studies of the same infectious disease in other populations often produced different and apparently conflicting results. Initially, it was believed that diverse HLA associations in different geographical areas might reflect linkage of a common susceptibility gene elsewhere in the MHC to different HLA marker alleles. However, now that it is clear that MHC class I and II proteins function as antigen-presenting molecules and that the HLA loci are the actual immune-response genes, this explanation appears inadequate. Quite reasonably, the claimed associations were treated with some scepticism and in 1987 Klein opined that there was no convincing evidence of MHC associations with infectious diseases of humans.

However, with the introduction of more precise molecular typing and the recognition that large studies were required to identify convincing associations, a clearer picture is beginning to emerge. The infectious disease that has been studied most frequently is leprosy. This reflects in part the intriguing variation in the human immune response to the bacillus, leading to polar clinical types, but also the ease of studying a chronic rather than an acute condition. Todd and colleagues (1990) reviewed 24 studies and performed a meta-analysis of the data sets. Although the validity of such pooling might be questioned, the overall conclusion was clear cut. In numerous studies there is a strong association between the HLA class II allele HLA-DR2 and leprosy (of any type). However, in many studies no such association was detected. This heterogeneity was statistically significant. Intriguingly, for another mycobacterial disease, tuberculosis, there are three recent studies identifying an association with the same allele, HLA-DR2 (Bothamley et al. 1989; Khomenko et al. 1990; Brahmajothi et al. 1991). Yet again, a larger number of earlier studies failed to identify this association.

In the last ten years there have also been numerous studies of the influence HLA may have on the risk of HIV infection and on the rate of progression of infected individuals to AIDS. Although these studies have included disappointingly few subjects, a picture of probable heterogeneity in HLA associations is again emerging. An association of the common caucasian haplotype HLA-A1-B8-DR3 with rapid disease progression has been seen in some studies, but in several others HLA-B35 appeared to be a significant risk factor (Scorza Smerali et al. 1986; Steel et al. 1988; Kaslow et al. 1990).

Studies of MHC associations with infectious disease in other species have been limited by a far less detailed characterization of the various alleles and a lack of reagents for their definition. However, in chickens, Briles et al.(1977) identified an association between an MHC haplotype and resistance to the herpes virus-induced atypical lymphoma, Marek's disease.

3. HLA STUDIES OF MALARIA IN WEST AFRICA

Our first study of HLA and malaria was of West African children in collaboration with the U.K. Medical Research Council Laboratories in The Gambia. During 1988–1990, over 600 children with severe malaria from the peri-urban Banjul population were studied and compared with 1400 other Gambians (Hill et al. 1991). Typing of HLA class I antigens showed that the allele HLA-B53 was underrepresented in the children with severe disease (16% of cases) compared to the non-malaria control groups (25% antigen frequency). Although this antigen was also less frequent amongst cases of mild or uncomplicated malaria, this difference was not statistically significant. Analysis of HLA class II region genes identified an association between a haplotype bearing the HLA-DR allele, HLA-DRB1*1302, and resistance to severe malarial anaemia. Cerebral malaria and severe malarial anaemia are the two major clinical manifestations of severe malaria in African children.

Further analysis of HLA class II region genes suggested that the HLA class II association is primarily due to the HLA-DRB1*1302 gene and not to allelic variation in neighbouring loci. Recently, Kwiatkowski and colleagues have analysed one of the HLA class III region genes, TNF, in these samples. It had been speculated that the HLA associations with malaria might relate to variation in this closely linked TNF gene because TNF itself has been implicated in the pathogenesis of severe malaria. However, polymorphisms identified in the TNF gene do not account for the HLA associations with malaria (McGuire et al. 1994).

From the odds ratios measured in this case-control study, it is possible to estimate the magnitude of protection associated with the alleles HLA-B53 and HLA-DRB1*1302 (Hill et al. 1991). These are in the region of a 40–50% reduction in risk of developing severe malaria. This is clearly smaller than the 90% protection associated with the carrier genotype for sickle haemoglobin. However, because these HLA alleles are much more prevalent in this population than haemoglobin S, a preventive fraction calculation (Bengtsson & Thomson 1981) indicates that at least as many cases of severe malaria may be being prevented by these HLA types being in the population as are prevented by the presence of sickle haemoglobin.

As well as identifying these associations with individual HLA alleles analysis of the overall effect of HLA variation on risk of severe malaria identified a probable influence of HLA class II variation on risk of developing severe malarial anaemia. Subsequently, a further small study of HLA antigens and malaria has been undertaken in a rural area of The Gambia (Bennet et al. 1993). Only mild malaria cases were studied and, as in the larger study, no clear association of individual alleles with this syndrome was identified. However, analysis of the combined effects of all class II haplotypes indicated a significant effect of this variation on risk of developing clinical malaria.

4. TEMPORAL VARIATION IN SELECTION?

HLA-B53 is found at highest frequencies in sub-Saharan African countries where malaria transmission is most intense. This suggests that this high frequency might result, at least in part, from natural selection by malaria. If we assume for the moment a constant selection pressure, it is possible to estimate how long it would take for this allele to rise in frequency from the degree of protection associated with it and the population mortality from malaria. To maintain the balanced haemoglobin S polymorphism at a carrier rate of 13% in The Gambia, the fitness of carriers must be 7% greater than that of individuals with normal haemoglobin (Cavalli–Sforza & Bodmer 1971). Taking this difference in fitness as an indicator of the historical death rate from malaria, an estimate compatible with contemporary surveys (Greenwood *et al.* 1987), it can be calculated that to reach the current antigen frequency of HLA-B53 (25%) from an initial allele frequency of, say, 0.001% would take about 7000 years (Nei 1987) with directional selection and a selection coefficient of 0.028 (0.07 × [1− Odds Ratio]). Because the allele is rare for most of this time, the estimates with dominance and semi-dominance are very similar. Starting at the 0.5% allele frequency which is found in parts of Europe today would take approximately 2500 years. Interestingly, these estimates are similar to the length of time that it was, until recently, believed that malaria had been a major cause of childhood mortality in Africa. It had been assumed that prior to the start of farming in Africa, about 4000–7000 years ago, the population density would have been insufficient to maintain significant malaria transmission (Bruce-Chwatt 1988).

However, this scenario now seems unlikely in the light of new estimates of the transmissibility of malaria (Gupta *et al.* 1994*b*). Because immunity to malaria is incomplete and exposure does not prevent reinfection, it is unlikely that the supply of susceptible individuals would run out. Thus, low population densities in hunter gatherer populations would probably not have ruled out a significant mortality from this pathogen. Hence the time depth of malarial selection in Africa could be much greater than suspected hitherto.

Recent data from our laboratory (Yates, Marsh, Newbold *et al.*, unpublished) on the selection of α thalassaemia by malaria in Africa are consistent with this greater time depth and suggest that malarial selection may have been important for tens of thousands of years. If this has been the case the tidy congruence of the estimate for the time depth of HLA-B53 selection and that of malarial selection is lost. The former estimate is the shorter and it would appear that the HLA-B53 frequency should now be much higher than is observed in Africa had there been a constant degree of selection.

It is worth emphasizing the limitations of these calculations. Our estimates of the original HLA-B53 frequency is very approximate and the measurement of 40% protection in the case-control study has 95% confidence intervals of 19%–57%. It may be the case that HLA-B53 frequencies are influenced by other infectious pathogens and this is ignored in the calculation. However, if malaria has indeed been a major selective force for many tens of thousands of years it appears unlikely that the protective effect of HLA-B53 that we measured in 1988–1990 can have been constant over this period.

5. GEOGRAPHICAL VARIATION IN SELECTION

The geographical distribution of HLA-B53 has been fairly well defined (summarized in Hill *et al.* 1991). It reaches its highest frequencies in west Africa with reports of antigen frequencies up to 40%. In east and central Africa it is also at high frequency of typically 10–20%. Further south in Zimbabwe and South Africa its frequency falls to less than 5%. In other regions it is generally uncommon or absent with frequencies of 1% or less. Thus, rather like the haemoglobin S allele, HLA-B53 is found at highest frequencies in the parts of Africa where malaria transmission is most intense. It is tempting to infer from this distribution that the protective effect of HLA-B53 against malaria may be fairly widespread and consistent within Africa. Following this line of argument one might propose that its rarity in south-east Asia, for instance, where malaria has selected many thalassaemia alleles, simply reflects the same stochastic processes that failed to lead to the selection of haemoglobin S in south-east Asia and Melanesia.

However, we have now had the opportunity to measure the current protective effect of HLA-B53 in another part of Africa. In collaboration with K. Marsh and R. Snow and others at the KEMRI-Wellcome Coastal Research Unit at Kilifi on the Kenyan coast we have been HLA-typing children from another large case-control study designed to measure risk factors for severe malaria in east African children. In Kilifi, HLA-B53 is found in 16% of the general population. Amongst more than 300 children with severe malaria an almost identical frequency of this allele was found (unpublished data). Hence, the protective association found in The Gambia is either absent in this coastal Kenyan region or at least appears to be considerably diminished.

Furthermore, detailed analysis of HLA-DR allelic variation in the Kenyan case-control study has provided further evidence for geographical variation in HLA associations with severe malaria. In The Gambia, HLA-DRB1*1302 was associated with about a 50% reduction in risk of severe malarial anaemia. In Kilifi, a significant protective association with a different HLA-DR type has recently been identified (unpublished data).

Hence, the results on two large African studies of severe malaria are reminiscent of the outcome of many smaller studies of leprosy, HIV infection and tuberculosis. Clear associations identified in some regions may not be found in other geographical areas.

6. MOLECULAR ANALYSIS OF HLA ASSOCIATIONS

This variation in HLA association with infection has important implications for the evolution and maintenance of HLA polymorphism in the human population. It also poses interesting questions as to the molecular basis of this variation which appears more marked than the fairly limited variation in HLA associations with autoimmune disorders.

Indeed, a molecular understanding of the nature of the interaction between parasite epitopes and HLA antigens associated with altered disease susceptibility would be valuable for addressing a further very controversial issue in evolutionary studies of MHC diversity (Hughes & Nei 1992; Hill *et al.* 1992*a*). With the increasing acceptance of the importance of parasite-driven selection in the maintenance of polymorphism in the MHC attention has focused on the type of selection involved. In particular, theoretical studies have addressed the question of whether overdominant selection or frequency-dependent selection may be better at maintaining diversity over long time periods. It is clear that in theory both types of selection could provide a powerful means of preserving polymorphism (Denniston & Crow 1990; Takahata & Nei 1990), although some authors strongly favour one type of selection. Unfortunately, there is almost no evidence from field studies to indicate which is actually operating. However, identification of the molecular mechanisms underlying some MHC associations with infectious pathogens should eventually facilitate an analysis of particular cases.

We have begun to analyse the mechanisms underlying the HLA associations observed with severe malaria in Africa by adopting an approach termed reverse immunogenetics. This utilizes recent information on the specificity of interactions between particular HLA molecules and their bound peptides (Falk et al. 1991; Hill et al. 1992b). For example, we found that peptides bound to HLA-B53 have a strong preference for proline at position 2 of their sequence and that at the carboxy-terminus of the peptide, usually position 9, a large hydrophobic amino acid is preferred. This allowed us to synthesize peptides corresponding to this 'motif' from the sequences of *P. falciparum* antigens known to be expressed during the liver-stage of infection. This stage of infection was chosen as the likely site of interaction between class I molecules and the parasite because infected red blood cells do not express HLA class I molecules. We tested these peptides for binding to HLA-B53 and then used the peptides which bound to identify peptide-specific cytotoxic T lymphocytes (CTL) in the blood of Gambians exposed to *P. falciparum* malaria. We have found that a single malaria peptide is the target of CTL restricted by HLA-B53. This peptide, termed ls6, is a fragment of the malaria antigen LSA-1 (liver-stage antigen-1) which is highly conserved between parasite strains.

Hence, the molecular basis of the protective association between HLA-B53 and severe malaria observed in The Gambia may be as follows. Cytotoxic T lymphocytes are induced by natural infection with sporozoites. In individuals with HLA-B53 these cells target the immunodominant peptide ls6 which is expressed on the surface of hepatocytes infected with a liver-stage parasite. In individuals with other HLA types CTL of other specificities are induced but, possibly because the ls6-B53 complex is more immunogenic than these other peptide-HLA complexes, their CTL are less efficient at lysing infected liver cells. Although this mechanism has not been demonstrated directly with human hepatocytes, in mouse models of malaria CTL have been shown to be protective and to kill infected hepatocytes.

Further work in The Gambia has now led to the identification of malaria epitopes for several further HLA class I molecules (Hill et al. 1992b and unpublished data). Some of these epitopes are

conserved, like ls6, but others show considerable polymorphism. In particular, a HLA-B35 epitope in the circumsporozoite protein (CSP) gene is located in a highly variable region of the molecule.

In principle, it is possible to extend this approach to the analysis of HLA class II associations with malaria. We have recently characterized the sequence features of peptides bound to the HLA-DRB1*1302 molecule associated with resistance to severe malarial anaemia in The Gambia. However, the motifs of peptides bound to HLA-DR molecules are less clear cut than those identified for HLA class I alleles. Hence, a very large number of peptides need to be synthesized to encompass the possible epitopes. This is problematic for a micro-organism as large as a malaria parasite but may be more feasible for viral infections such as hepatitis B, for which a protective HLA-DR association has recently been identified (M. Thursz, H. C. Thomas, A. V. S. Hill et al., unpublished data). An alternative, more challenging option, is to sequence directly individual parasite peptides eluted from the HLA-DR molecules of infected cells. The amounts of these peptides are too small to analyse by conventional sequencers, but the use of new tandem mass spectrometers may make this feasible in the future.

7. WHY MAY HLA ASSOCIATIONS VARY?

The analysis of epitopes for HLA class I molecules in *P. falciparum* suggests one explanation for variation in HLA associations with malaria. There is polymorphism in epitopes in the circumsporozoite protein for some HLA class I molecules including HLA-B35 and HLA-B7. The HLA-B35 epitope has been analysed in some detail. There are at least four variants of this epitope found amongst *P. falciparum* parasites in The Gambia (table 1). Two of these, cp26 and cp29 have been shown to be epitopes for HLA-B35 restricted CTL (Hill et al. 1992b). Further analysis has shown that cp27 and cp28 fail to bind to HLA-B35, explaining their lack of recognition by the CTL. It is tempting to speculate that the polymorphism observed in this region of CSP has been driven by immune pressure from HLA-B35-restricted CTL with cp27 and cp28 as escape mutants. HLA-B35 is the commonest HLA class I molecule in The Gambia, so that this observation could be extended further to invoke a form of frequency-dependent selection for particular parasite alleles of CSP.

Although parasite epitopes for the HLA-DRB1*302 allele have not yet been identified, it is likely that this allele's association with resistance to severe malarial anaemia is related to an enhanced immune response to a blood-stage *P. falciparum* antigen. The major antigens of the blood stage parasite are characterized by extensive polymorphism, and much of this may relate to immune selection pressures. Furthermore, there is now geographical information on the distribution of some of these parasite variants (Creasey et al. 1990). A particular allele of the major merozoite surface antigen, MSA-1, is predominant in The Gambia (Conway & McBride 1991). In contrast, in Kilifi, this allele has recently been found to be present at a much lower frequency (S. Kyes, C. I. N. Newbold, K. Marsh *et al.*, personal communication). Hence, geographical variation in the frequency of parasite alleles could be an important determinant of variable HLA associations.

We have recently analysed possible sequence variation in the HLA-B53 epitope, ls6, amongst parasites in The Gambia and in Kilifi, Kenya. In both areas, no sequence variation was found in the region of the LSA-1 gene that encodes this epitope. Hence, variability of this peptide does not underlie the differential association of this HLA-B53 with resistance to severe malaria in these two areas. However, polymorphism in CTL epitopes may still be relevant to this association. In case-control studies only the *relative* protective efficacies of HLA alleles can be assessed so that a less-effective series of other HLA alleles in The Gambia would give rise to greater apparent protection associated with HLA-B53 in The Gambia than in Kenya. The diminished efficacy of the other HLA types in The Gambia might relate to polymorphism in the parasite gene encoding their particular class I epitopes.

An alternative explanation for geographical variation in HLA associations with malaria relates to differences in transmission intensities in different areas. The intensity of malaria transmission in Kilifi is probably several-fold higher than in the periurban area of The Gambia. Two further features of malaria have been incorporated into a model of the effects of CTL and other interventions on the prevalence of severe and mild malaria in a population (Gupta *et al.*, unpublished). The first is the assumption that there are important differences in virulence between various strains of *P. falciparum* so that common mild strains result in episodes of mild malaria and rare severe strains lead to the rarer severe (cerebral) malaria. The second feature of this model is that blood-stage immunity to disease in malaria is strain-specific.

Table 1. *Amino acid sequences of the four allelic peptides: cp26–cp29*

(These form part of a variable region of the circumsporozite protein of *Plasmodium falciparum* (amino acids 368–375). The sequences are shown using the standard one letter code for amino acids. These four peptides are present in different parasite strains found in The Gambia. Cp26 and cp29 are epitopes for cytotoxic T lymphocytes restricted by the HLA class I molecule HLA-B35. Cp27 and cp28 fail to bind to HLA-B35 and are not epitopes. Cytotoxic T lymphocytes specific for cp26 have been shown to fail to recognize the other three variants. Similarly, cytotoxic T lymphocytes specific for cp29 fail to recognize cp26 – 28.)

	1	2	3	4	5	6	7	8
cp26	K	P	K	D	E	L	D	Y
cp27	K	P	K	D	Q	L	D	Y
cp28	K	P	K	D	Q	L	N	Y
cp29	K	S	K	D	E	L	D	Y

There is evidence to support both of these assumptions.

Figure 1 shows the relationship between the cumulative probability of disease and the number of times an individual has encountered a particular strain of *P. falciparum*. The non-linearity reflects the reduced probability of disease upon encountering a strain that has been met many times before, a consequence of strain-specific blood-stage immunity. The key feature of this model is that interventions at an early stage of infection, e.g. against the liver stage parasite via CTL or against the mosquito by the use of impregnated bed nets (Gupta *et al.* 1994*a*), have a disproportionately greater effect on the incidence of severe malaria than mild malaria. Hence, this type of model may explain the observed greater reduction in severe than mild malaria with bed nets (Alonso *et al.* 1993), and the stronger association of HLA-B53 with resistance to severe than mild malaria (Hill *et al.* 1991). Secondly, in areas of higher transmission, where children have experienced more infections, the slope of the line (figure 1) will be flatter than in a lower-transmission area such as The Gambia. Also, in higher transmission areas, many of the otherwise severe cases of malaria will happen during infancy while there is still protection from maternal antibodies or other protective mechanisms during the early months of life. Both of these factors will have the effect of attenuating differences in the protective efficacy of differential levels of CTL response, and diminish HLA class I associations. A further prediction of the model, that the protective HLA-B53 association in The Gambia should have been more evident in

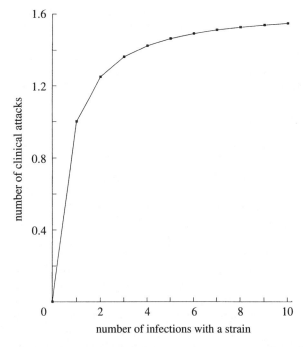

Figure 1. The relationship between the average number of clinical attacks of malaria caused by a single strain of *Plasmodium falciparum* during a year in a given individual and the number of infections caused by that strain. This non-linear relationship reflects an increasing degree of strain-specific anti-disease immunity to a parasite strain with increasing exposure to it.

younger than older children was indeed observed (unpublished data).

Hence, HLA associations with malaria may be sensitive to the transmission intensity of this infection. It can be seen that in this case the outcome is a function of particular features of the infection such as the presence of immune responses to more than one stage of the life-cycle of the malaria parasite. However, more generally, a multiplicity of different protective immune responses is a feature of host defence against many natural pathogens. It seems likely that the most important protective response may often vary according the host genotype and also the parasite strain. Hence, just as class I and class II restricted responses to the liver stage and blood stage of malaria respectively, may be of varying importance in different geographical areas, if the host uses difference immune clearance mechanisms during the same stage of infection, according to transmission patterns in the locality or according to local parasite genotype, again HLA associations may reflect this heterogeneity.

It is likely that as particular immune defence mechanisms are understood in more detail, we shall develop a clearer picture of how sensitive HLA associations may be to variation in environmental factors and parasite genotype.

8. FLUCTUATING SELECTION

As mentioned above, overdominant selection and frequency-dependent selection are both powerful means by which genetic polymorphism may be maintained in a population. Although it seems likely that both will prove to be important in maintaining such diversity, we shall consider here a third potential mechanism. Forty years ago Levene (1953) and Dempster (1955) found that spatially varying selection could maintain polymorphism without (arithmetic-mean) overdominance, and in 1963 Haldane & Jayakar demonstrated that temporally varying selection coefficients could lead to a permanent polymorphism. This happens provided that the geometric mean fitness of the heterozygote is highest over time. Hedrick (1974), Gillespie (1978, 1985) and Takahata (1981) have developed more detailed models of fluctuating selection incorporating finite population size, mutation and genetic drift.

The general conclusion is that fluctuating selection can lead either to allele loss or preservation of diversity depending on the precise circumstances. Unfortunately, no theoretical work appears to have addressed fluctuating selection and the special circumstances of MHC diversity. It seems likely that conditions may frequently exist in which the 'holding power' (Karlin & Lieberman 1974; Takahata 1981; Gillespie 1985) of fluctuating selection outweighs the increased tendency to lose rare alleles (Karlin & Lieberman 1974; Takahata 1981; Gillespie 1985). There are a large number of fairly frequent HLA alleles in most populations so that the tendency to lose very rare ones may be less relevant, and cyclic variation, which may be more effective than random changes in

selection coefficients in maintaining variation (Hedrick 1974), is probably a common feature of host–parasite interactions.

Analysis of fluctuating selection must be particularly relevant to parasite-driven selection and HLA variation. Infectious pathogens frequently appear in epidemics, often of novel antigenic type, so temporally varying selection coefficients are likely to be common. Extensive antigenic diversity exists in the majority of natural human pathogens that have been studied in detail, and there is increasing evidence of geographical variation in parasite allele frequencies. Certain pathogens, such as HIV, also undergo extensive mutation within a single host. Finally, as discussed above, several environmental and genetic factors can lead to different HLA associations with a single infectious pathogen in particular regions.

9. CONCLUSIONS

It is becoming clear that HLA associations with infectious pathogens such as *P. falciparum* exist, but that large studies using molecular HLA typing are required to demonstrate these convincingly. Some of the apparently contradictory results in earlier studies of HLA and infectious diseases may be accounted for by real heterogeneity in HLA associations between different populations. Recently, the use of new molecular immunological techniques has made it possible to begin to dissect the molecular mechanisms underlying these associations and, in turn, such work has revealed possible reasons for heterogeneity in associations. Irrespective of the underlying mechanisms, the observations of variable HLA associations with infection in field studies highlight the relevance of fluctuating selection pressures to the evolutionary maintenance of diversity in the MHC. Further theoretical and experimental studies of fluctuating selection pressures should assist in assessing their importance in maintaining MHC polymorphism.

REFERENCES

Alonso, P.L., Lindsay, S.W, Armstrong Shellenberg, J.R.M. *et al.* 1993 A malaria control trial using insecticide-treated bed nets and targeted chemoprophylaxis in a rural area of The Gambia, West Africa. 6.The impact of the interventions on mortality and morbidity from malaria. *Trans. R. Soc. trop. Med. Hyg.* **87**, Suppl. 2, 37–44.

Bengtsson, B.O. & Thomson, G. 1981 Measuring the strength of associations between HLA antigens and diseases. *Tissue Antigens* **18**, 356–363.

Bennett, S., Allen, S.J., Olerup, O. *et al.* 1993 Human leucocyte antigen and malaria morbidity in a Gambian population. *Trans. R. Soc. trop. Med. Hyg.* **87**, 286–287.

Bothamley, G.H., Beck, J.S., Schreuder, G.M.T. *et al.* 1989 Association of tuberculosis and Mycobacterium tuberculosis specific antibody levels with HLA. *J. Infect. Dis.* **59**, 549–555.

Brahmajothi, V., Pitchappan, R.M., Kakkanaiah, V.N. *et al.* 1991 Association of pulmonary tuberculosis and HLA in South India. *Tubercle* **72**, 123–132.

Briles, W.E., Stone, H.A. & Cole, R.K. 1977 Marek's disease: effects of B histocompatibility alloalleles in resistant and susceptible chickens. *Science* **195**, 193–195.

Bruce–Chwatt, L.J. 1988 In *Malaria: principles and practice of malariology.* (ed. W. H. Wernsdorfer & I. A. McGregor), pp. 159. Edinburgh: Churchill-Livingstone.

Cavalli–Sforza, L.L. & Bodmer, W.F. 1971 *The genetics of human populations*, pp. 133–184. San Francisco: W. H. Freeman Co.

Conway, D.J. & McBride J.S. 1991 Population genetics of *Plasmodium falciparum* in a malaria hyperendemic area. *Parasitology* **103**, 7–16.

Creasey, A., Fenton, B., Walker, A. *et al.* 1990 Genetic diversity of *Plasmodium falciparum* shows geographical variation. *Am. J. trop. Med. Parasitol.* **42**, 403–413.

Dempster, E.R. 1955 Maintenance of genetic heterogeneity. *Cold Spring Harb. Symp. quant. Biol.* **20**, 25–32.

Denniston, C. & Crow, J.F. 1990 Alternative fitness models with the same allele frequency dynamics. *Genetics* **125**, 201–205.

Doherty, P.C. & Zinkernagel, R.M. 1975 A biological role for the major histocompatibility antigens *Lancet* i, 1406–1409.

Falk, K., Roetzschke, O., Stevanovic, S., Jung, G. & Rammensee, H.-G. 1991 Allele-specific motifs revealed by sequencing of self-peptides eluted from MHC molecules. *Nature* **351**, 290–296.

Gillespie, J.H. 1978 A general model to account for enzyme variation in natural populations. V. The SAS-CFF model. *Theor. Popul. Biol.* **14**, 1–45.

Gillespie, J.H. 1985 The interaction of genetic drift and mutation with selection in a fluctuating environment. *Theor. Popul. Biol.* **27**, 222–237.

Greenwood, B.M., Bradley, A.K., Greenwood, A.M. *et al.* 1987 Mortality and morbidity from malaria among children in a rural area of The Gambia, West Africa. *Trans. R. Soc. trop. Med. Hyg.* **81**, 478–486.

Gupta, S., Hill, A.V.S., Kwiatkowski, D., Greenwood, A.M., Greenwood, B.M. & Day, K.P. 1994 Parasite virulence and disease patterns in *Plasmodium falciparum* malaria. *Proc. natn. Acad. Sci. U.S.A.* **91**, 3715–3719.

Gupta, S., Trenholme, K., Anderson, R.M. & Day, K.P. 1994b Antigenic diversity and the transmission dynamics of *Plasmodium falciparum*. *Science* **263**, 961–963.

Haldane, J.B.S. and Jayakar, S.D. 1963 Polymorphism due to selection of varying direction. *J. Genet.* **58**, 237–242.

Hedrick, P.W. 1974 Genetic variation in a heterogeneous environment. I. Temporal heterogeneity and the absolute dominance model. *Genetics* **78**, 757–770.

Hill, A.V.S., Allsopp, C.E.M., Kwiatkowski, D. *et al.* 1991 Common West African HLA antigens are associated with protection from severe malaria. *Nature* **352**, 595–600.

Hill, A.V.S., Kwiatkowski, D., McMichael, A.J., Greenwood, B.M. & Bennett, S. 1992a Maintenance of major histocompatibility complex polymorphism. *Nature* **355**, 403.

Hill A.V.S., Elvin J., Willis A. *et al.* 1992b Molecular analysis of the association of HLA-B53 and resistance to severe malaria. *Nature* **360**, 434–439.

Hughes, A.L. & Nei, M. 1992 Maintenance of MHC polymorphism. *Nature* **355**, 402–403.

Karlin, S. & Lieberman, U. 1974 Temporal fluctuations in selection intensities: case of large population size. *Theor. Popul. Biol.* **6**, 355–382.

Kaslow R.A., Duquesnoy R., Van Raden M. *et al.* 1990 A1, Cw7, B8, DR3 HLA antigen combination associated with rapid decline of T-helper lymphocytes in HIV-1 infection. *Lancet* **335**, 927–930.

Khomenko, A.G., Litvinov, V.I, Chukanova, V.P. & Pospelov, L.E. 1990 Tuberculosis in patients with various HLA phenotypes. *Tubercle* **71**, 187–192.

Klein, J. 1987 Origin of major histocompatibility complex polymorphism: the trans-species hypothesis *Hum. Immun.* **19**, 155–162.

Levene, H. 1953 Genetic equilibrium when more than one ecological niche is available. *Am. Nat.* **87**, 331–333.

McGuire, W., Hill, A.V.S., Allsopp, C.E.M., Greenwood, B.M. & Kwiatkowski, D. 1994 Variation in the TNF-α promoter region associated with susceptibility to cerebral malaria. *Nature, Lond.* **371**, 508–511.

Nei, M. 1987 *Molecular evolutionary genetics*, pp. 335–339. New York: Columbia University Press.

Scorza Smeraldi, R., Fabio, G., Lazzarin, A. *et al.* 1986 HLA-associated susceptibility to acquired immunodeficiency syndrome in Italian patients with HIV infection. *Lancet* ii, 1187–1189.

Snell, G.D. 1968 The H-2 locus of the mouse: observations and speculations concerning its comparative genetics and its polymorphism. *Folia biol.* **14**, 335–358.

Steel, C.M., Ludlam, C.A., Beatson, D. *et al.* 1988 HLA haplotype A1 B8 DR3 as a risk factor for HIV-related disease. *Lancet* i, 1185–1188.

Takahata, N. 1981 Genetic variability and rate of gene substitution in a finite population under mutation and fluctuating selection. *Genetics* **98**, 427–440.

Takahata, N. & Nei, M. 1990 Alleleic genealogy under overdominant and frequency-dependent selection and polymorphism of the major histocompatibility complex. *Genetics* **124**, 967–978.

Todd, J.R., West, B.C. & McDonald, J.C. 1990 Human leukocyte antigen and leprosy: study in northern Louisiana and review. *Rev. Infect. Dis.* **12**, 63–74.

Index